MATLAB® Software for the Code Excited Linear Prediction Algorithm

The Federal Standard–1016

Synthesis Lectures on Algorithms and Software in Engineering

Editor
Andreas S. Spanias, *Arizona State University*

MATLAB® Software for the Code Excited Linear Prediction Algorithm – The Federal Standard–1016
Karthikeyan N. Ramamurthy and Andreas S. Spanias
2009

Advances in Modern Blind Signal Separation Algorithms: Theory and Applications
Kostas Kokkinakis and Philipos C. Loizou
2010

OFDM Systems for Wireless Communications
Adarsh B. Narasimhamurthy, Mahesh K. Banavar, and Cihan Tepedelenlioğlu
2010

Algorithms and Software for Predictive Coding of Speech
Atti Venkatraman
2010

Advances in Waveform-Agile Sensing for Tracking
Sandeep Prasad Sira, Antonia Papandreou-Suppappola, and Darryl Morrell
2008

Despeckle Filtering Algorithms and Software for Ultrasound Imaging
Christos P. Loizou and Constantinos S. Pattichis
2008

MATLAB® Software for the Code Excited Linear Prediction Algorithm – The Federal Standard–1016

Karthikeyan N. Ramamurthy and Andreas S. Spanias

ISBN: 978-3-031-00386-8   paperback
ISBN: 978-3-031-01514-4   ebook

DOI 10.1007/978-3-031-01514-4

A Publication in the Springer series
SYNTHESIS LECTURES ON ALGORITHMS AND SOFTWARE IN ENGINEERING

Lecture #3
Series Editor: Andreas S. Spanias, *Arizona State University*
Series ISSN
Synthesis Lectures on Algorithms and Software in Engineering
Print 1938-1727   Electronic 1938-1735

MATLAB® Software for the Code Excited Linear Prediction Algorithm

The Federal Standard–1016

Karthikeyan N. Ramamurthy and Andreas S. Spanias
Arizona State University

SYNTHESIS LECTURES ON ALGORITHMS AND SOFTWARE IN ENGINEERING #3

ABSTRACT

This book describes several modules of the Code Excited Linear Prediction (CELP) algorithm. The authors use the Federal Standard-1016 CELP MATLAB® software to describe in detail several functions and parameter computations associated with analysis-by-synthesis linear prediction. The book begins with a description of the basics of linear prediction followed by an overview of the FS-1016 CELP algorithm. Subsequent chapters describe the various modules of the CELP algorithm in detail. In each chapter, an overall functional description of CELP modules is provided along with detailed illustrations of their MATLAB® implementation. Several code examples and plots are provided to highlight some of the key CELP concepts.

The MATLAB® code within this book can be found at http://www.morganclaypool.com/page/fs1016

KEYWORDS

speech coding, linear prediction, CELP, vocoder, ITU, G.7XX standards, secure communications

Contents

Preface

The CELP algorithm, proposed in the mid 1980's for analysis-by-synthesis linear predictive coding, had a significant impact in speech coding applications. The CELP algorithm still forms the core of many speech coding standards that exist nowadays. Federal Standard-1016 is an early standardized version of the CELP algorithm and is based on the enhanced version of the originally proposed AT&T Bell Labs CELP coder [1].

This book presents MATLAB®software that simulates the FS-1016 CELP algorithm. We describe the theory and implementation of the algorithm along with several detailed illustrations using MATLAB®functions. In the first chapter, we introduce the theory behind speech analysis and linear prediction and also provide an overview of speech coding standards. From Chapter 2 and on, we describe the various modules of the FS-1016 CELP algorithm. The theory sections in most chapters are self-sufficient in that they explain the necessary concepts required to understand the algorithm. The implementation details and illustrations with MATLAB®functions complement the theoretical underpinnings of CELP. Furthermore, the MATLAB®programs and the supporting files are available for download in the companion website of the book. The extensive list of references allows the reader to carry out a more detailed study on specific aspects of the algorithm and they cover some of the recent advancements in speech coding.

We note that although this is an existing algorithm and perhaps the very first that has been standardized, it is a great starting point for learning the core concepts associated with CELP. Students will find it easy to understand the CELP functions by using the software and the associated simulations. Practitioners and algorithm developers will be able to understand the basic concepts and develop several new functions to improve the algorithm or to adapt it for certain new applications. We hope this book will be useful in understanding the CELP algorithm as well as the more recent speech coding standards that are based on CELP principles.

The authors acknowledge Ted Painter, who wrote the MATLAB®code for the FS-1016 CELP algorithm, and DoD for developing the C code for the algorithm.

The MATLAB® code within this book can be found at http://www.morganclaypool.com/page/fs1016

Karthikeyan N. Ramamurthy and Andreas S. Spanias
February 2010

CHAPTER 1

Introduction to Linear Predictive Coding

This book introduces linear predictive coding and describes several modules of the Code Excited Linear Prediction (CELP) algorithm in detail. The MATLAB program for Federal Standard-1016 (FS-1016) CELP algorithm is used to illustrate the components of the algorithm. Theoretical explanation and mathematical details along with relevant examples that reinforce the concepts are also provided in the chapters. In this chapter, some of the basics of linear prediction starting from the principles of speech analysis are introduced along with some fundamental concepts of speech analysis.

Linear prediction analysis of speech signals is at the core of most narrowband speech coding standards. A simple engineering synthesis model that has been used in several speech processing and coding applications is shown in Figure 1.1. This source-system model is inspired by the human speech production mechanism. Voiced speech is produced by exciting the vocal tract filter with quasi-periodic glottal pulses. The periodicity of voiced speech is due to the vibrating vocal chords. Unvoiced speech is produced by forcing air through a constriction in the vocal tract or by using other mechanisms that do not engage the vocal chords.

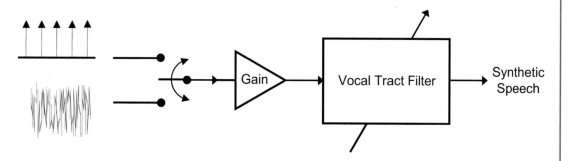

Figure 1.1: Engineering model for speech synthesis.

The vocal tract is usually represented by a tenth-order digital all-pole filter. As shown in Figure 1.1, voiced speech is produced by exciting the vocal tract filter with periodic impulses and unvoiced speech is generated using random pseudo-white noise excitation. Filter coefficients and excitation parameters are typically determined every 20 ms or less for speech sampled at 8 kHz. The filter frequency response models the formant structure of speech and captures the resonant modes of the vocal tract.

1.1   LINEAR PREDICTIVE CODING

Digital filters for linear predictive coding applications are characterized by the difference equation,

$$y(n) = b(0)x(n) - \sum_{i=1}^{M} a(i)y(n-i) \,. \tag{1.1}$$

In the input-output difference equation above, the output $y(n)$ is given as the sum of the input minus a linear combination of past outputs (feedback term). The parameters $a(i)$ and $b(i)$ are the filter coefficients or filter taps and they control the frequency response characteristics of the filter. Filter coefficients are programmable and can be made adaptive (time-varying). A direct-form realization of the digital filter is shown in Figure 1.2.

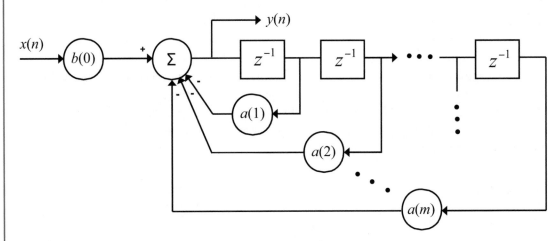

Figure 1.2: Linear prediction synthesis filter.

When the tenth-order all-pole filter representing the vocal tract is excited by white noise, the signal model corresponds to an autoregressive (AR) time-series representation. The coefficients of the AR model can be determined using linear prediction techniques. The application of linear prediction in speech processing, and specifically in speech coding, is often referred to as Linear Predictive Coding (LPC). The LPC parameterization is a central component of many compression algorithms that are used in cellular telephony for bandwidth compression and enhanced privacy. Bandwidth is conserved by reducing the data rate required to represent the speech signal. This data rate reduction is achieved by parameterizing speech in terms of the AR or all-pole filter coefficients and a small set of excitation parameters. The two excitation parameters for the synthesis configuration shown in Figure 1.1 are the following: (a) the voicing decision (voiced/unvoiced) and (b) the pitch period. In the simplest case, 160 samples (20 ms for 8 kHz sampling) of speech can be represented with ten all-pole vocal tract parameters and two parameters that specify the excitation signal. Therefore, in this case, 160 speech samples can be represented by only twelve parameters which results in a data compression ratio of more than 13 to 1 in terms of the number of parameters that will be encoded and transmitted. In new standardized algorithms, more elaborate forms of parameterization exploit further the redundancy in the signal and yield better compression and much improved speech quality.

1.1.1   VOCAL TRACT PARAMETER ESTIMATION

Linear prediction is used to estimate the vocal tract parameters. As the name implies, the linear predictor estimates the current sample of the speech signal using a linear combination of the past samples. For example, in a tenth-order linear predictor, an estimate of the current speech sample $s(n)$ is produced using a linear combination of the ten previous samples, i.e., $s(n-1), s(n-2), \ldots, s(n-10)$. This is done by forming a prediction error,

$$e(n) = s(n) - \sum_{i=1}^{10} a(i) s(n-i) .$$ (1.2)

The prediction parameters $a(i)$ are unknown and are determined by minimizing the Mean-Square-Error (MSE) $E[e^2(n)]$. The prediction parameters $a(i)$ are also used to form the all-pole digital filter for speech synthesis. The minimization of the MSE yields a set of autocorrelation equations that can be represented in terms of the matrix equation,

$$\begin{bmatrix} r(1) \\ r(2) \\ r(3) \\ \cdot \\ \cdot \\ r(10) \end{bmatrix} = \begin{bmatrix} r(0) & r(1) & r(2) & \ldots & r(9) \\ r(1) & r(0) & r(1) & \ldots & r(8) \\ r(2) & r(1) & r(0) & \ldots & r(7) \\ \cdot & \cdot & \cdot & \ldots & \cdot \\ \cdot & \cdot & \cdot & \ldots & \cdot \\ r(9) & r(8) & r(7) & \ldots & r(0) \end{bmatrix} \begin{bmatrix} a(1) \\ a(2) \\ a(3) \\ \cdot \\ \cdot \\ a(10) \end{bmatrix} .$$ (1.3)

The autocorrelation sequence can be estimated using the equation,

$$r(m) = \frac{1}{N} \sum_{n=0}^{N-m-1} s(n+m) s(n) .$$ (1.4)

The integer N is the number of speech samples in the frame (typically $N = 160$). The autocorrelations are computed once per speech frame and the linear prediction coefficients are computed by inverting the autocorrelation matrix, i.e.,

$$
\begin{bmatrix} a(1) \\ a(2) \\ a(3) \\ . \\ . \\ a(10) \end{bmatrix} = \begin{bmatrix} r(0) & r(1) & r(2) & \cdots & r(9) \\ r(1) & r(0) & r(1) & \cdots & r(8) \\ r(2) & r(1) & r(0) & \cdots & r(7) \\ . & . & . & \cdots & . \\ . & . & . & \cdots & . \\ r(9) & r(8) & r(7) & \cdots & r(0) \end{bmatrix}^{-1} \begin{bmatrix} r(1) \\ r(2) \\ r(3) \\ . \\ . \\ r(10) \end{bmatrix}. \tag{1.5}
$$

The coefficients $a(1), a(2), \ldots, a(10)$ form the transfer function in (1.6) of the filter which is used in the speech synthesis system shown in Figure 1.1. This all-pole filter reproduces speech segments using either random (unvoiced) or periodic (voiced) excitation.

$$
H(z) = \frac{1}{1 - \displaystyle\sum_{i=1}^{10} a(i)z^{-i}}. \tag{1.6}
$$

The matrix in (1.3) can be inverted efficiently using the Levinson-Durbin order-recursive algorithm given in (1.7 a-d).

for $m = 0$, initialize:
$\varepsilon_0^f = r(0)$
   for $m = 1$ to 10:

$$
a_m(m) = \frac{r(m) - \displaystyle\sum_{i=1}^{m-1} a_{m-1}(i)\, r(m-i)}{\varepsilon_{m-1}^f} \tag{1.7 a-d}
$$

   for $i = 1$ to $m - 1$:
      $a_m(i) = a_{m-1}(i) - a_m(m)\, a_{m-1}(m-i), \quad 1 \leq i \leq m-1$
   end
$\varepsilon_m^f = \left(1 - (a_m(m))^2\right) \varepsilon_{m-1}^f$
End

The subscript m of the prediction parameter $a_m(i)$ in the recursive expression (1.7 a-d) is the order of prediction during the recursion while the integer in the parenthesis represents the coefficient index. The symbol ε_m^f represents the mean-square estimate of the prediction residual during the recursion. The coefficients $a_m(m)$ are called *reflection* coefficients and are represented by the symbol $k_m = a_m(m)$. In the speech processing literature, the negated reflection coefficients are also known as Partial Correlation (*PARCOR*) coefficients. Reflection coefficients correspond to *lattice* filter structures that have been shown to be electrical equivalents of acoustical models of the vocal tract. Reflection coefficients have good quantization properties and have been used in several speech compression systems for encoding vocal tract parameters. A lattice structure is shown in Figure 1.3.

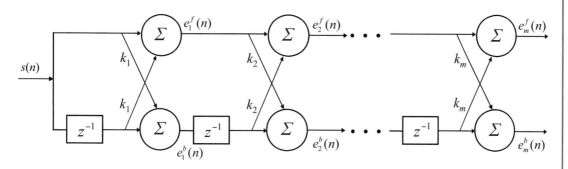

Figure 1.3: Lattice structure for linear prediction and reflection coefficients.

1.1.2 EXCITATION, GAIN AND PITCH PERIOD

The residual signal $e(n)$ given in (1.2) forms the optimal excitation for the linear predictor. For low-rate representations of speech, the residual signal is typically replaced by a parametric excitation signal model. Many of the early algorithms for linear prediction used the two-state excitation model (impulses/noise) shown in Figure 1.1. This model is parameterized in terms of the gain, the binary voicing parameter and the pitch period. The gain of voiced and unvoiced segments is generally determined such that the short-term energy of the synthetic speech segment matches that of the analysis segment. For unvoiced speech, the excitation is produced by a random number generator. Since unvoiced segments are associated with small energy and large number of zero crossings, voicing can be determined by energy and zero-crossing measurements. Pitch estimation and tracking is a difficult problem. Some of the well-known algorithms for pitch detection were developed in the late sixties and seventies. A straightforward approach to pitch detection is based on selecting the peak of the autocorrelation sequence (excluding $r(0)$). A more expensive but also more robust pitch detector relies on peak-picking the *cepstrum*. The Simplified Inverse Filter Tracking (SIFT) algorithm is based on peak-picking the autocorrelation sequence of the prediction residual associated with downsampled speech. Post-processing algorithms for pitch smoothing are also used to provide frame-to-frame pitch continuity. Current pitch detection algorithms yield high-resolution (sub-sample) estimates for the pitch period and are often specific to the analysis-synthesis system. In many Analysis-by-Synthesis (A-by-S) linear predictive coders, the pitch is measured by a closed-loop process which accounts for the impact of the pitch on the overall quality of the reconstructed speech.

1.1.3   LINEAR PREDICTION PARAMETER TRANSFORMATIONS

One of the major issues in LPC is the quantization of the linear prediction parameters. Quantization of the direct-form coefficients $a(i)$ is generally avoided. The reflection coefficients k_m are by-products of the Levinson algorithm and are more robust for quantization. The reflection coefficients can also be quantized in an ordered manner, i.e., the first few reflection coefficients can be encoded with a higher precision. Transformations of the reflection coefficients, such as the Log Area Ratio (LAR), i.e.,

$$\text{LAR}(m) = \log \left\{ \frac{1 + k_m}{1 - k_m} \right\}, \tag{1.8}$$

have also been used in several LPC algorithms. Another representation of Linear Prediction (LP) parameters that is being used in several standardized algorithms consists of Line Spectrum Pairs (LSPs). In LSP representations, a typical tenth-order polynomial,

$$A(z) = 1 + a_1 z^{-1} + \cdots + a_{10} z^{-10}, \tag{1.9}$$

is represented by the two auxiliary polynomials $P(z)$ and $Q(z)$ where,

$$P(z) = A(z) + z^{-11} A\left(z^{-1}\right), \tag{1.10}$$

$$Q(z) = A(z) - z^{-11} A\left(z^{-1}\right). \tag{1.11}$$

$P(z)$ and $Q(z)$ have a set of five complex conjugate pairs of zeros each that typically lie on the unit circle. Hence, each polynomial can be represented by the five frequencies of their zeros (the other five frequencies are their negatives). These frequencies are called Line Spectral Frequencies (LSFs). If the polynomial $A(z)$ is minimum phase, the roots of the polynomials $P(z)$ and $Q(z)$ alternate on the unit circle.

1.1.4   LONG-TERM PREDICTION

Almost all modern LP algorithms include long-term prediction in addition to the tenth-order short-term linear predictor. Long Term Prediction (LTP) captures the long-term correlation in the speech signal and provides a mechanism for representing the periodicity of speech. As such, it represents the *fine* harmonic structure in the short-term speech spectrum. The LTP requires estimation of two parameters, i.e., a delay τ and a gain parameter $a(\tau)$. For strongly voiced segments, the delay is usually an integer that approximates the pitch period. The transfer function of a simple LTP synthesis filter is given by,

$$A_\tau(z) = \frac{1}{1 - a(\tau)z^{-\tau}}. \tag{1.12}$$

The gain is obtained by the equation $a(\tau) = r(\tau)/r(0)$. Estimates of the LTP parameters can be obtained by searching the autocorrelation sequence or by using closed-loop searches where the LTP lag that produces the best speech waveform matching is chosen.

1.2   ANALYSIS-BY-SYNTHESIS LINEAR PREDICTION

In closed-loop source-system coders, shown in Figure 1.4, the excitation source is determined by closed-loop or A-by-S optimization. The optimization process determines an excitation sequence that minimizes the perceptually-weighted MSE between the input speech and reconstructed speech [1, 2, 3]. The closed-loop LP combines the spectral modeling properties of vocoders with the waveform matching attributes of waveform coders; and hence, the A-by-S LP coders are also called *hybrid* LP coders. The term hybrid is used because of the fact that A-by-S integrates vocoder and waveform coder principles. The system consists of a short-term LP synthesis filter $1/A(z)$ and a LTP synthesis filter $1/A_L(z)$ as shown in Figure 1.4. The Perceptual Weighting Filter (PWF) $W(z)$ shapes the error such that quantization noise is masked by the high-energy formants. The PWF is given by,

$$W(z) = \frac{A(z/\gamma_1)}{A(z/\gamma_2)} = \frac{1 - \sum_{i=1}^{m} \gamma_1(i)\, a(i)\, z^{-i}}{1 - \sum_{i=1}^{m} \gamma_2(i)\, a(i)\, z^{-i}} ; \quad 0 < \gamma_2 < \gamma_1 < 1 , \qquad (1.13)$$

where γ_1 and γ_2 are the adaptive weights, and m is the order of the linear predictor. Typically, γ_1 ranges from 0.94 to 0.98; and γ_2 varies between 0.4 and 0.7 depending upon the tilt or the flatness characteristics associated with the LPC spectral envelope [4, 5]. The role of $W(z)$ is to de-emphasize the error energy in the formant regions [6]. This de-emphasis strategy is based on the fact that, in the formant regions, quantization noise is partially masked by speech. From Figure 1.4, note that a gain factor g scales the excitation vector $\mathbf{x}$ and the excitation samples are filtered by the long-term and short-term synthesis filters.

The three most common excitation models typically embedded in the excitation generator module (Figure 1.4) in the A-by-S LP schemes include: the *Multi-Pulse Excitation* (MPE) [2, 3], the *Regular Pulse Excitation* (RPE) [7], and the vector or *Code Excited Linear Prediction* (CELP) [1]. A 9.6 kb/s Multi-Pulse Excited Linear Prediction (MPE-LP) algorithm is used in *Skyphone* airline applications [8]. A 13 kb/s coding scheme that uses RPE [7] was adopted for the full-rate ETSI GSM Pan-European digital cellular standard [9]. The standard was eventually replaced by the GSM Enhanced Full-Rate (EFR) described briefly later.

The aforementioned MPE-LP and RPE schemes achieve high-quality speech at medium rates. For low-rate high-quality speech coding a more efficient representation of the excitation sequence is required. Atal [10] suggested that high-quality speech at low rates may be produced by using non-instantaneous (delayed decision) coding of Gaussian excitation sequences in conjunction with A-by-S linear prediction and perceptual weighting. In the mid-eighties, Atal and Schroeder [1, 11] proposed a CELP algorithm for A-by-S linear predictive coding.

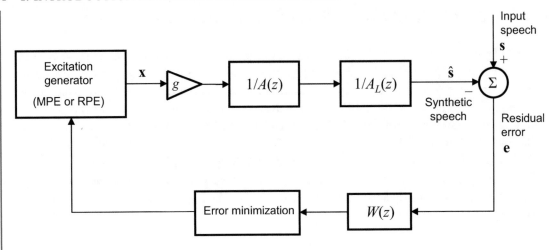

Figure 1.4: A typical source-system model employed in the analysis-by-synthesis LP.

The excitation codebook search process in CELP can be explained by considering the A-by-S scheme shown in Figure 1.5. The $N \times 1$ error vector $\mathbf{e}$ associated with the ith excitation vector, can be written as,

$$\mathbf{e}^{(i)} = \mathbf{s}_w - \hat{\mathbf{s}}_w^0 - g_k \hat{\mathbf{s}}_w^{(i)} , \tag{1.14}$$

where $\mathbf{s}_w$ is the $N \times 1$ vector that contains the perceptually-weighted speech samples, $\hat{\mathbf{s}}_w^0$ is the vector that contains the output due to the initial filter state, $\hat{\mathbf{s}}_w^{(i)}$ is the filtered synthetic speech vector associated with the ith excitation vector, and $g^{(i)}$ is the gain factor. Minimizing $\zeta^{(i)} = \mathbf{e}^{(i)^T} \mathbf{e}^{(i)}$ with respect to $g^{(i)}$, we obtain,

$$g^{(i)} = \frac{\bar{\mathbf{s}}_w^T \hat{\mathbf{s}}_w^{(i)}}{\hat{\mathbf{s}}_w^{(i)^T} \hat{\mathbf{s}}_w^{(i)}} , \tag{1.15}$$

where $\bar{\mathbf{s}}_w = \mathbf{s}_w - \mathbf{s}_w^0$, and T represents the transpose of a vector. From (1.15), $\zeta^{(i)}$ can be written as,

$$\zeta^{(i)} = \bar{\mathbf{s}}_w^T \bar{\mathbf{s}}_w - \frac{\left(\bar{\mathbf{s}}_w^T \hat{\mathbf{s}}_w^{(i)}\right)^2}{\hat{\mathbf{s}}_w^{(i)^T} \hat{\mathbf{s}}_w^{(i)}} . \tag{1.16}$$

The ith excitation vector $\mathbf{x}^{(i)}$ that minimizes (1.16) is selected and the corresponding gain factor $g^{(i)}$ is obtained from (1.15). Note that the perceptual weighting, $W(z)$, is applied directly on the input speech, $\mathbf{s}$, and synthetic speech, $\hat{\mathbf{s}}$, in order to facilitate for the CELP analysis that follows. The codebook index, i, and the gain, $g^{(i)}$, associated with the candidate excitation vector, $\mathbf{x}^{(i)}$, are encoded and transmitted along with the short-term and long-term prediction filter parameters.

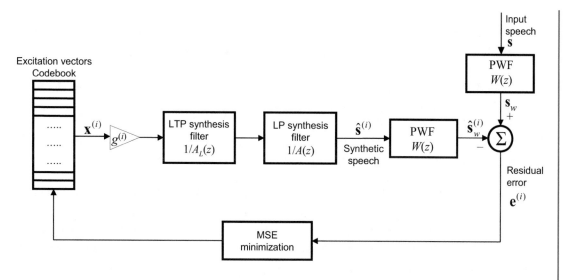

Figure 1.5: A generic block diagram for the A-by-S Code Excited Linear Predictive (CELP) coding.

One of the disadvantages of the original CELP algorithm is the large computational complexity required for the codebook search [1]. This problem motivated a great deal of work focused upon developing structured codebooks [12, 13] and fast search procedures [14]. In particular, Davidson and Gersho [12] proposed sparse codebooks and Kleijn *et al.* [13] proposed a fast algorithm for searching stochastic codebooks with overlapping vectors. In addition, Gerson and Jasiuk [15, 16] proposed a Vector Sum Excited Linear Predictive (VSELP) coder which is associated with fast codebook search and robustness to channel errors. Other implementation issues associated with CELP include the quantization of the CELP parameters, the effects of channel errors on CELP coders, and the operation of the algorithm on finite-precision and fixed-point machines. A study on the effects of parameter quantization on the performance of CELP was presented in [17], and the issues associated with the channel coding of the CELP parameters were discussed by Kleijn in [18]. Some of the problems associated with the fixed-point implementation of CELP algorithms were presented in [19].

1.2.1 CODE EXCITED LINEAR PREDICTION ALGORITHMS

In this section, we taxonomize CELP algorithms into three categories that are consistent with the chronology of their development, i.e., first-generation CELP (1986-1992), second-generation CELP (1993-1998), and third-generation CELP (1999-present).

1.2.1.1 First-Generation CELP Coders

The first-generation CELP algorithms operate at bit rates between 4.8 kb/s and 16 kb/s. These are generally high complexity and non-toll quality algorithms. Some of the first-generation CELP algorithms include the following: the FS-1016 CELP, the IS-54 VSELP, the ITU-T G.728 Low-Delay (LD) CELP, and the IS-96 Qualcomm CELP. The FS-1016 4.8 kb/s CELP standard [20, 21] was jointly developed by the Department of Defense (DoD) and the Bell Labs for possible use in the third-generation Secure Telephone Unit (STU-III). The IS-54 VSELP algorithm [15, 22] and its variants are embedded in three digital cellular standards, i.e., the 8 kb/s TIA IS-54 [22], the 6.3 kb/s Japanese standard [23], and the 5.6 kb/s half-rate GSM [24]. The VSELP algorithm uses highly structured codebooks that are tailored for reduced computational complexity and increased robustness to channel errors. The ITU-T G.728 Low-Delay (LD) CELP coder [25, 26] achieves low one-way delay by using very short frames, a backward-adaptive predictor, and short excitation vectors (5 samples). The IS-96 Qualcomm CELP (QCELP) [27] is a variable bit rate algorithm and is part of the Code Division Multiple Access (CDMA) standard for cellular communications. Most of these standardized CELP algorithms were eventually replaced by newer second and third generation A-by-S coders.

1.2.1.2 Second-Generation Near-Toll-Quality CELP Coders

The second-generation CELP algorithms are targeted for Internet audio streaming, Voice-over-Internet-Protocol (VoIP), teleconferencing applications, and secure communications. Some of the second-generation CELP algorithms include the following: the ITU-T G.723.1 dual-rate speech codec [28], the GSM EFR [23, 29], the IS-127 Relaxed CELP (RCELP) [30, 31], and the ITU-T G.729 Conjugate Structured - Algebraic CELP (CS-ACELP) [4, 32].

The coding gain improvements in second-generation CELP coders can be attributed partly to the use of algebraic codebooks in excitation coding [4, 32, 33, 34]. The term Algebraic CELP (ACELP) refers to the structure of the codebooks used to select the excitation codevector. Various algebraic codebook structures have been proposed [33, 35], but the most popular is the *interleaved pulse permutation* code. In this codebook, the codevector consists of a set of interleaved permutation codes containing only few non-zero elements. This is given by,

$$p_i = i + jd, \quad j = 0, 1, \ldots, 2^M - 1, \tag{1.17}$$

where p_i is the pulse position, i is the pulse number, and d is the interleaving depth. The integer M is the number of bits describing the pulse positions. Table 1.1 shows an example ACELP codebook structure, where, the interleaving depth, $d = 5$, the number of pulses or tracks are equal to 5, and the number of bits to represent the pulse positions, $M = 3$. From (1.17), $p_i = i + j5$, where $i = 0, 1, 2, 3, 4$ and $j = 0, 1, 2, \ldots, 7$.

Track(i)	Pulse positions (P_i)
0	P_0: 0, 5, 10, 15, 20, 25, 30, 35
1	P_1: 1, 6, 11, 16, 21, 26, 31, 36
2	P_2: 2, 7, 12, 17, 22, 27, 32, 37
3	P_3: 3, 8, 13, 18, 23, 28, 33, 38
4	P_4: 4, 9, 14, 19, 24, 29, 34, 39

Table 1.1: An example algebraic codebook structure; tracks and pulse positions.

For a given value of i, the set defined by (1.17) is known as 'track,' and the value of j defines the pulse position. From the codebook structure shown in Table 1.1, the codevector, $x(n)$, is given by,

$$x(n) = \sum_{i=0}^{4} \alpha_i \delta(n - p_i), \quad n = 0, 1, \ldots, 39, \tag{1.18}$$

where $\delta(n)$ is the unit impulse, α_i are the pulse amplitudes (± 1), and p_i are the pulse positions. In particular, the codebook vector, $x(n)$, is computed by placing the 5 unit pulses at the determined locations, p_i, multiplied with their signs (± 1). The pulse position indices and the signs are encoded and transmitted. Note that the algebraic codebooks do not require any storage.

1.2.1.3 Third-Generation (3G) CELP for 3G Cellular Standards

The 3G CELP algorithms are multimodal and accommodate several different bit rates. This is consistent with the vision on wideband wireless standards [36] that will operate in different modes: low-mobility, high-mobility, indoor, etc. There are at least two algorithms that have been developed and standardized for these applications. In Europe, GSM standardized the Adaptive Multi-Rate (AMR) coder [37, 38], and in the U.S., the Telecommunications Industry Association (TIA) has tested the Selectable Mode Vocoder (SMV) [39, 40, 41]. In particular, the adaptive GSM multirate coder [37, 38] has been adopted by European Telecommunications Standards Institute (ETSI) for GSM telephony. This is an ACELP algorithm that operates at multiple rates: 12.2, 10.2, 7.95, 6.7, 5.9, 5.15, and 5.75 kb/s. The bit rate is adjusted according to the traffic conditions.

The Adaptive Multi-Rate WideBand (AMR-WB) [5] is an ITU-T wideband standard that has been jointly standardized with 3GPP, and it operates at rates 23.85, 23.05, 19.85, 18.25, 15.85, 14.25, 12.65, 8.85 and 6.6 kbps. The higher rates provide better quality where background noise is stronger. The other two important speech coding standards are the Variable Rate Multimode WideBand (VMR-WB) and the extended AMR-WB (AMR-WB+). The VMR-WB has been adopted as the new 3GPP2 standard for wideband speech telephony, streaming and multimedia messaging systems [42]. VMR-WB is a Variable Bit Rate (VBR) coder with the source controlling the bit rate of operation: Full-Rate (FR), Half-Rate (HR), Quarter-Rate (QR) or Eighth-Rate

(ER) encoding to provide the best subjective quality at a particular Average Bit Rate (ABR). The source-coding bit rates are 13.3, 6.2, 2.7 and 1 kbps for FR, HR, QR and ER encoding schemes. The AMR-WB+ [43] can address mixed speech and audio content and can consistently deliver high quality audio even at low bit rates. It has been selected as an audio coding standard in 2004 by ETSI and 3rd Generation Partnership Project (3GPP). The SMV algorithm (IS-893) was developed to provide higher quality, flexibility, and capacity over the existing IS-96 QCELP and IS-127 Enhanced Variable Rate Coding (EVRC) CDMA algorithms. The SMV is based on four codecs: full-rate at 8.5 kb/s, half-rate at 4 kb/s, quarter-rate at 2 kb/s, and eighth-rate at 0.8 kb/s. The rate and mode selections in SMV are based on the frame voicing characteristics and the network conditions. G.729.1 is a recent speech codec adopted by ITU-T that can handle both narrowband and wideband speech and is compatible with the widely used G.729 codec [44]. The codec can operate from 8 kb/s to 32 kb/s with 8 kb/s and 12 kb/s in the narrowband and 14 kb/s to 32 kb/s at 2 kb/s intervals in the wideband. The codec incorporates bandwidth extension and provides bandwidth and bit rate scalability at the same time.

1.2.2   THE FEDERAL STANDARD-1016 CELP

A 4.8 kb/s CELP algorithm has been adopted in the late 1980s by the DoD for use in the STU-III. This algorithm is described in the Federal Standard-1016 (FS-1016) [21] and was jointly developed by the DoD and AT&T Bell labs. Although new algorithms for use with the STU emerged, such as the Mixed Excitation Linear Predictor (MELP), the CELP FS-1016 remains interesting for our study as it contains core elements of A-by-S algorithms that are still very useful. The candidate algorithms and the selection process for the standard are described in [45]. The synthesis configuration for the FS-1016 CELP is shown in Figure 1.6. Speech in the FS-1016 CELP is sampled at 8 kHz and segmented in frames of 30 ms duration. Each frame is segmented into sub-frames of 7.5 ms duration. The excitation in this CELP is formed by combining vectors from an adaptive and a stochastic codebook with gains g_a and g_s, respectively (gain-shape vector quantization). The excitation vectors are selected in every sub-frame by minimizing the perceptually-weighted error measure. The codebooks are searched sequentially starting with the adaptive codebook. The term "adaptive codebook" is used because the LTP lag search can be viewed as an adaptive codebook search where the codebook is defined by previous excitation sequences (LTP state) and the lag τ determines the specific vector. The adaptive codebook contains the history of past excitation signals and the LTP lag search is carried over 128 integer (20 to 147) and 128 non-integer delays. A subset of lags is searched in even sub-frames to reduce the computational complexity. The stochastic codebook contains 512 sparse and overlapping codevectors [18]. Each codevector consists of sixty samples and each sample is ternary valued $(1, 0, -1)$ [46] to allow for fast convolution.

Ten short-term prediction parameters are encoded as LSPs on a frame-by-frame basis. Sub-frame LSPs are obtained by applying linear interpolation of frame LSPs. A short-term pole-zero postfilter (similar to that proposed in [47]) is also part of the standard. The details on the bit allocations are given in the standard. The computational complexity of the FS-1016 CELP was

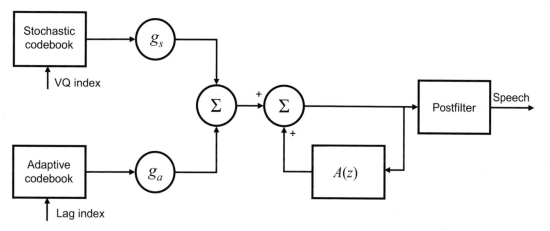

Figure 1.6: FS-1016 CELP synthesis.

estimated at 16 Million Instructions per Second (MIPS) for partially searched codebooks and the Diagnostic Rhyme Test (DRT) and Mean Opinion Scores (MOS) were reported to be 91.5 and 3.2, respectively.

1.3   SUMMARY

This chapter introduced the basics of linear predictive coding, A-by-S linear prediction and provided a review of the speech coding algorithms based on CELP. A detailed review of the basics of speech coding, algorithms and standards until the early 1990's can be found in [48]. A more recent and short review of the modern speech coding algorithms is available in the book by Spanias *et al.* [49]. From Chapter 2 and on, we will describe the details of the FS-1016 standard and provide MATLAB program for all the pertinent functions. Chapter 2 will describe the autocorrelation analysis and linear prediction module starting from the framing of speech signal upto the computation of reflection coefficients from the bandwidth expanded LP coefficients. The construction of LSP polynomials, computation of LSFs and their quantization are described in Chapter 3. Various distance measures that are used to compute spectral distortion are detailed in Chapter 4. Chapter 5 illustrates the adaptive and stochastic codebook search procedures, and the final chapter describes the components in the FS-1016 decoder module.

CHAPTER 2

Autocorrelation Analysis and Linear Prediction

Linear Prediction (LP) analysis is performed on the framed and windowed input speech to compute the direct-form linear prediction coefficients. It involves performing autocorrelation analysis of speech and using the Levinson-Durbin algorithm to compute the linear prediction coefficients. The autocorrelation coefficients are calculated and high frequency correction is done to prevent ill conditioning of the autocorrelation matrix. The autocorrelation lags are converted to LP coefficients using Levinson-Durbin recursion. Then, the direct-form LP coefficients are subjected to bandwidth expansion. The bandwidth expanded LP coefficients are converted back to Reflection Coefficients (RCs) using the inverse Levinson-Durbin recursion. The block diagram in Figure 2.1 illustrates the autocorrelation analysis and linear prediction module of the FS-1016 CELP transmitter. Every chapter starting from this one, will have a similar schematic block diagram that will illustrate the overall function of the module described in that chapter. Each block in the diagram corresponds to a particular script in the FS-1016 MATLAB code, whose name is provided in the block itself.

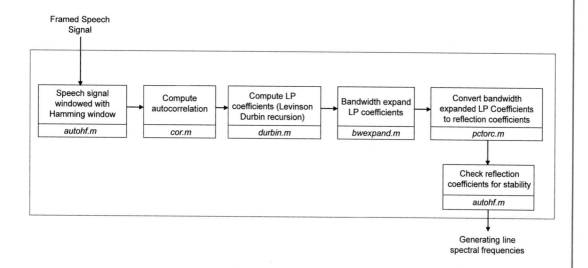

Figure 2.1: Autocorrelation analysis and linear prediction.

The following sections of this chapter demonstrate the computation of autocorrelation lags, conversion of the autocorrelation lags to LP coefficients and recursive conversion of the LP coefficients to RCs using MATLAB programs.

2.1   FRAMING AND WINDOWING THE INPUT SPEECH

The speech frame used in the FS-1016 transmitter is of size 240 samples and corresponds to 30 ms of speech at the standardized sampling rate of 8000 Hz. There are three speech buffers used in the encoder. Two speech buffers, s_{old} and s_{new} are one frame behind and one frame ahead, respectively, while the sub-frame analysis buffer, s_{sub}, is half frame behind s_{new} and half frame ahead of s_{old}. In the later part of the FS-1016 analysis stage, s_{sub} will be divided into four sub-frames each of size 60 samples corresponding to 7.5 ms of speech. The input speech buffer, s_{new}, is scaled and clamped to 16 bit integer range, before processing. Clamping to 16 bit integer range is performed by thresholding the entries of the speech buffer. It is then filtered using a high-pass filter to remove low frequency noise and windowed with a Hamming window for high frequency correction. Program P2.1 loads an input speech frame and creates the three speech frame buffers in the variables snew, sold and ssub.

Example 2.1   The Hamming window used to window the speech signal is shown in Figure 2.2. The input speech (s_{new}) and the windowed speech (s) frames are shown in Figures 2.3 and 2.4,

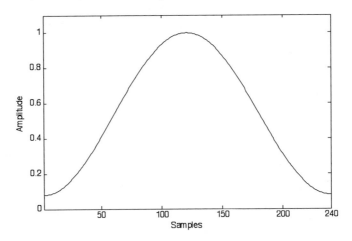

Figure 2.2: The 240 point hamming window.

respectively, and they are generated using Program P2.1. The plot of the Fourier transforms for the input and the windowed speech frames are provided in Figures 2.5 and 2.6. The Fourier transform of the windowed speech frame is smoother because of the convolution of the frequency response of the Hamming window in the frequency domain.

```
% P2.1  - frame_window.m

function [s,snew] = frame_window(iarf,w,sold)
% sold  - Initialized to zeros for the first frame
% iarf  - Input speech frame, 240 samples
% scale - Scaling parameter, initialized to 1
% maxval- Maximum possible value, 32767
% minval- Minimum possible value, -32768
% ll    - Length of buffer, 240 samples
% w     - Hamming Window Sequence
load framew_par.mat;
load hpf_param.mat;

ssub=zeros(240,1);

% SCALE AND CLAMP INPUT DATA TO 16-BIT INTEGER RANGE
snew = min( [ (iarf .* scale)'; maxval ] )';
snew = max( [ snew'; minval ] )';

% RUN HIGHPASS FILTER TO ELIMINATE HUM AND LF NOISE
[ snew, dhpf1 ] = filter( bhpf, ahpf, snew, dhpf1 );

% MAINTAIN SSUB SUBFRAME ANALYSIS BUFFER.  IT IS
% 1/2 FRAME BEHIND SNEW AND 1/2 FRAME AHEAD OF SOLD.
ssub( 1:ll/2 ) = sold( (ll/2)+1:ll );
ssub( (ll/2)+1:ll ) = snew( 1:ll/2 );

% UPDATE SOLD WITH CONTENTS OF SNEW
sold = snew;

% APPLY WINDOW TO INPUT SEQUENCE
s = snew.* w;
```

Program P2.1: Framing and windowing.

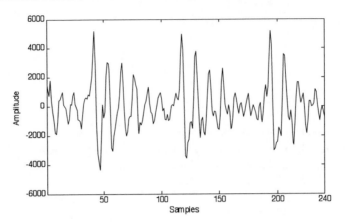

Figure 2.3: Input speech frame.

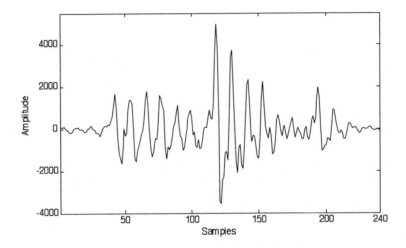

Figure 2.4: Windowed speech frame.

2.2   COMPUTATION OF AUTOCORRELATION LAGS

The biased autocorrelation estimates are calculated using the modified time average formula [50],

$$r(i) = \sum_{n=1+i}^{N} u(n)u(n-i), \qquad i = 0, 1, \ldots, M , \qquad (2.1)$$

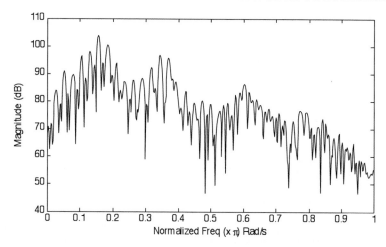

Figure 2.5: Fourier transform of input speech frame.

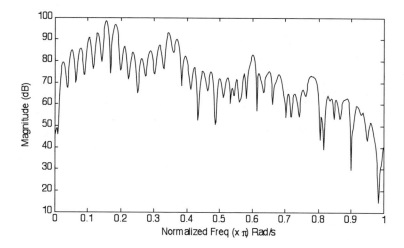

Figure 2.6: Fourier transform of windowed speech frame.

where, r is the autocorrelation value that is computed, i is the lag, u is the input signal and N is the length of the input signal. Program P2.2 computes the autocorrelation lags of a test signal u for the maximum lag specified by M.

Example 2.2   A sinusoidal test signal (u) shown in Figure 2.7 is used to estimate autocorrelation using Program P2.2, and the autocorrelation estimates are computed for lags 0 to 20 as shown in

```
% P2.2 - cor.m

function [r0 r] = cor(u,M)
% u - Input signal.
% N - Length of input signal.
% M - Maximum lag value.
N=length(u);

% ALLOCATE RETURN VECTOR
r = zeros( M, 1 );
% COMPUTE r(0)
r0 = sum( u(1:N) .* u(1:N) );
% COMPUTE C(i), i NONZERO
for i = 1:M
    r(i) = sum( u(i+1:N) .* u(1:N-i) );
end
```

Program P2.2: Computation of autocorrelation lags.

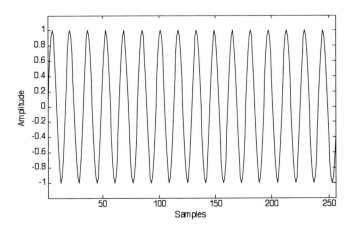

Figure 2.7: Sinusoidal test signal.

Figure 2.8. The sinusoidal test signal has a period of 16 samples, and it can be readily observed from the autocorrelation function, which has a peak at a lag of 16.

2.3   THE LEVINSON-DURBIN RECURSION

Linear predictor coefficients are the coefficients of the LP polynomial $A(z)$. Recall that LP analysis is carried out using the all-zero filter $A(z)$ whose input is the preprocessed speech frame and the output is the LP residual. But, during synthesis the filter used has the transfer function $1/A(z)$, which models the vocal tract and reproduces the speech signal when excited by the LP residual.

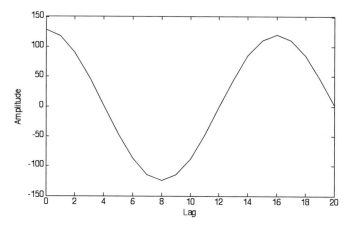

Figure 2.8: The autocorrelation estimate for the sinusoidal test signal.

The Levinson-Durbin recursion is used to compute the coefficients of the LP analysis filter from the autocorrelation estimates [51]. It exploits the Toeplitz structure of the autocorrelation matrix to recursively compute the coefficients of a tenth-order filter.

The algorithm [52] initializes the MSE as follows,

$$\varepsilon_0 = r(0) \,, \tag{2.2}$$

where, ε_0 is the initial MSE and $r(0)$ is the autocorrelation. Then, for a first-order predictor,

$$a_1(1) = \frac{-r(1)}{r(0)} \,, \tag{2.3}$$

$$k_1 = a_1(1) \,, \tag{2.4}$$

$$q = r(1) \,, \tag{2.5}$$

where, $a_m(i)$ is the ith predictor coefficient of order m and k_m is the reflection coefficient.

We find the linear predictor coefficients recursively using (2.6) to (2.9). Note that, in the MATLAB code given in Program P2.3, the index of linear predictor coefficients start from 1, whereas in the equations the index starts from 0. For $i = 1$ to $m-1$,

```
% P2.3  - durbin.m

function a = durbin( r0, r, m )
% r0 - Zero lag autocorrelation(value)
% r  - Autocorrelation for lags 1 to m+1
% m  - Predictor Order
% a  - LP coefficients (initialized as column vector, size m+1, a(1)=1)

% INITIALIZATION
a=zeros(m+1,1);
a(1)=1;
E = r0;
a(2) = -r(1) / r0;
k(1) = a(2);
q = r(1);

% RECURSION
for i = 1:m-1
    E= E + ( q * k(i) );
    q = r( i+1 );
    q = q + sum( r( 1:i ) .* a( i+1:-1:2 ) );
    k( i+1 ) = -q / E;
    tmp( 1:i ) = k( i+1 ) .* a( i+1:-1:2 );
    a( 2:i+1 ) = a( 2:i+1 ) + tmp( 1:i )';
    a( i+2 ) = k( i+1 );
end
```

Program P2.3: Levinson-Durbin recursion.

$$\varepsilon_i = \varepsilon_{i-1} + k_i q \,, \tag{2.6}$$

$$q = r(i+1) + \sum_{j=1}^{i} a_i(i+1-j)r(i) \,, \tag{2.7}$$

$$k_{i+1} = \frac{-q}{\varepsilon_i} \,, \tag{2.8}$$

$$\varepsilon_i = \varepsilon_{i-1}(1 - k_i^2) \,, \tag{2.9}$$

$$a_{i+1}(j) = a_i(j) + k_{i+1}a_i(i+1-j) \,. \tag{2.10}$$

Example 2.3   The frequency response magnitude of the LP synthesis filter $1/A(z)$, where the coefficients of $A(z)$ are calculated using the Program P2.3, is shown in Figure 2.9. The autocorrelation estimates (r) of the speech signal (s) shown in Figure 2.4 are used to compute the coefficients of $A(z)$.

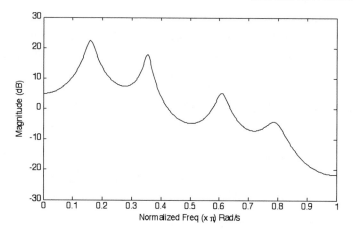

Figure 2.9: Frequency response of the linear prediction synthesis filter.

2.4   BANDWIDTH EXPANSION

Bandwidth expansion by a factor of γ changes the linear predictor coefficients from $a(i)$ to $\gamma^i a(i)$ and moves the poles of the LP synthesis filter inwards. The bandwidth expansion can be computed by [53],

$$BW = 2\cos^{-1}\left[1 - \frac{(1 - \gamma)^2}{2\gamma} \right], \qquad (2.11)$$

where $3 - 2\sqrt{2} \leq \gamma \leq 1$, and BW is the bandwidth expansion in radians. At a sampling frequency of 8000 Hz, a bandwidth expansion factor of 0.994127 produces a bandwidth expansion of 15 Hz. This bandwidth expansion factor is used in the implementation of the FS-1016 algorithm. Bandwidth expansion expands the bandwidth of the formants and makes the transfer function somewhat less sensitive to round-off errors.

```
% P2.4 - bwexp.m

function aexp = bwexp( gamma, a, m )
% gamma - Bandwidth expansion factor, 0.99412
% a - Linear predictor coefficients, column vector
% aexp - Bandwidth expanded linear predictor coefficients
% m - Order of the linear predictor

% SCALE PREDICTOR COEFFICIENTS TO SHIFT POLES RADIALLY INWARD
aexp( 1:m+1, 1 ) = ( a( 1:m+1 ) .* ( gamma .^ (0:m)' ) );
```

Program P2.4: Bandwidth expansion.

Example 2.4   The pole-zero plots of the original and bandwidth-expanded filter coefficients for a tenth-order linear predictor are shown in Figure 2.10. The corresponding frequency response is shown in Figure 2.11. The linear prediction filter coefficients are calculated for the speech signal given in Figure 2.4. The filter coefficients, before and after bandwidth expansion by a factor of 0.994127 are given in Table 2.1. Though not quite evident in this case, after bandwidth expansion, the poles have moved inwards in the pole-zero plot.

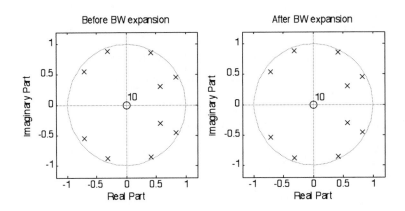

Figure 2.10: Pole-zero plots of linear prediction filter.

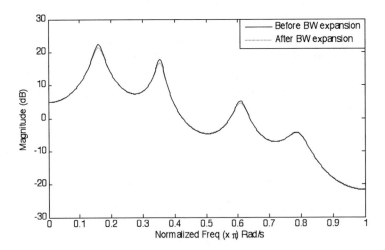

Figure 2.11: Frequency response before and after bandwidth expansion.

Coefficient	Before Bandwidth Expansion	After Bandwidth Expansion
a_0	1.0000	1.0000
a_1	-1.6015	-1.5921
a_2	1.5888	1.5702
a_3	-1.3100	-1.2870
a_4	1.4591	1.4251
a_5	-1.3485	-1.3094
a_6	1.0937	1.0558
a_7	-0.7761	-0.7447
a_8	0.9838	0.9385
a_9	-0.7789	-0.7386
a_{10}	0.2588	0.2440

Table 2.1: Linear Prediction Filter coefficients.

In order to demonstrate bandwidth expansion more clearly, let us take an example of the Long Term Predictor (LTP), where the filter expression is given by $1/(1 - 0.9z^{-10})$. The bandwidth expansion factor γ is taken to be 0.9. Please note that this is only an instructive example and bandwidth expansion is generally not applied to the LTP. The pole-zero plot is given in Figure 2.12 and the frequency response is given in Figure 2.14. In the pole-zero plot, it is quite evident that the poles have moved towards the origin and the frequency response also clearly shows expanded bandwidth. The plots given in Figure 2.13 show that the impulse response after bandwidth expansion decays faster than the one before bandwidth expansion.

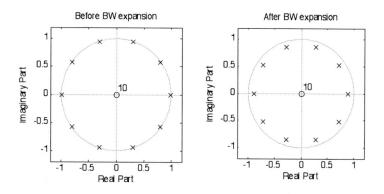

Figure 2.12: Pole-Zero plots of linear prediction filter.

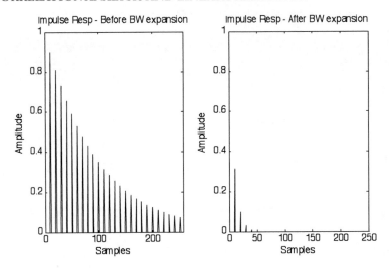

Figure 2.13: Impulse response before and after bandwidth expansion.

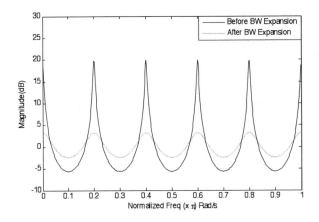

Figure 2.14: Frequency response before and after bandwidth expansion.

2.5   INVERSE LEVINSON-DURBIN RECURSION

Inverse Levinson-Durbin recursion is used to compute the RCs from the bandwidth expanded linear predictor coefficients [50]. Here, the coefficients are real-valued, therefore the order-update equations for the forward and backward prediction filters are given in a matrix format by,

$$
\begin{bmatrix} a_i(j) \\ a_i(i-j) \end{bmatrix} = \begin{bmatrix} 1 & k_i \\ k_i & 1 \end{bmatrix} \begin{bmatrix} a_{i-1}(j) \\ a_{i-1}(i-j) \end{bmatrix}, \qquad j = 0, 1, \ldots, i \,, \tag{2.12}
$$

where the order i varies from 1 to m. Solving for $a_{i-1}(j)$ from the above equation, we get,

$$a_{i-1}(j) = \frac{a_i(j) - k_i a_i(i - j)}{1 - |k_i|^2}, \qquad j = 0, 1, \dots, i, \qquad (2.13)$$

where, k_i is the reflection coefficient of order i. Starting from the highest order m, we compute the linear predictor coefficients for orders decreasing from $m - 1$ through 1 using (2.13). Then, using the fact that $k_i = a_i(i)$, we determine the RCs of orders 1 through $m - 1$. If the absolute values of all the RCs are less than 1, then the LP polynomial is minimum phase resulting in a stable LP synthesis filter. This can be intuitively understood from (2.9), where if $|k_i| < 1$, $\varepsilon_i < \varepsilon_{i-1}$ which means that the MSE decreases with the order of recursion as we expect for a stable filter. If $|k_i| = 1$ or $|k_i| > 1$ the MSE is either zero or it increases with the order of recursion, which is not a characteristic of a stable filter. Detailed explanation of the relationship between the stability of the LP synthesis filter and the magnitude of reflection coefficients is given in [54].

In the MATLAB code given in Program P2.5, there are certain variable and indexing modifications that are specific to our implementation. First, we take $k_i = -a_i(i)$ because of the sign convention that forces k_1 to be positive for voiced speech. Because of this, when (2.12) is implemented in the program, the numerator is a summation instead of difference. The indices for the array start from 1, and vectorized operations are used to compute the coefficient array in a single statement.

```
% P2.5 - pctorc.m

function k = pctorc( lpc, m )
% MAXNO - Predictor Order (10 here)
% lpc   - LP Coefficients
% m     - Order of LP polynomial
MAXNO=m;

% ALLOCATE RETURN VECTOR AND INIT LOCAL COPY OF PREDICTOR POLYNOMIAL
k = zeros( MAXNO, 1 );
a = lpc;
% DO INVERSE LEVINSON DURBIN RECURSION
for i = m:-1:2
    k(i) = -a( i+1 );
    t( i:-1:2 ) = ( a( i:-1:2 ) + ( k(i) .* a(2:i) ) ) / ...
                  ( 1.0 - ( k(i) * k(i) ) );
    a( 2:i ) = t( 2:i );
end
k(1) = -a(2);
```

Program P2.5: Inverse Levinson-Durbin recursion.

Example 2.5   In order to keep the examples simple, we work with a reduced order LP polynomial. This is achieved by taking only the first four autocorrelation lags (r0_4, r_4) of the speech frame

(s) and forming a fourth order LP polynomial using Levinson-Durbin recursion. The coefficients of the fourth order LP polynomial are bandwidth-expanded (a_4exp) and provided as input to the Program P2.5 to yield the reflection RCs (k_4). The coefficients are given as vectors below. Note that the vectors are given in MATLAB notation where separation using semicolon indicates a column vector. In this case, the LP synthesis filter is stable because the absolute values of all the reflection coefficients are less than 1.

$$\text{a\_4exp} = [1.0000; -1.5419; 1.3523; -0.7197; 0.3206],$$
$$\text{k\_4} = [0.7555; -0.7012; 0.2511; -0.3206].$$

2.6 SUMMARY

LP analysis is performed on the speech frames and the direct-form LP coefficients are obtained using the autocorrelation analysis and linear prediction module. Further information on linear prediction is available in a tutorial on linear prediction by Makhoul [55] and the book by Markel and Gray [56]. The government standard FS-1015 algorithm is primarily based on open-loop linear prediction and is a precursor of the A-by-S FS-1016 [57]. The LP coefficients are bandwidth-expanded and then converted to RCs. The negative of the RCs ($-k_i$) are also known as PARCOR coefficients [58], and they can be determined using the harmonic analysis algorithm proposed by Burg [59]. Note that the stability of the LP synthesis filter can be verified by checking the magnitude of RCs. RCs are less sensitive to round-off noise and quantization errors than the direct-form LP coefficients [60]. Because of the good quantization properties, RCs have been used in several of the first generation speech coding algorithms. Second and third generation coders, however, use Line Spectrum Pairs (LSPs) that are described in the next chapter.

CHAPTER 3

Line Spectral Frequency Computation

Line Spectrum Pairs (LSPs) are an alternative LP spectral representation of speech frames that have been found to be perceptually meaningful in coding systems. LSPs can be quantized using perceptual criteria and have good interpolation properties. Two LSP polynomials can be formed from the LP polynomial $A(z)$. When $A(z)$ is minimum phase, the zeros of the LSP polynomial have two interesting properties: (1) they lie on the unit circle and (2) the zeros of the two polynomials are interlaced [61]. Each zero corresponds to a LSP frequency and instead of quantizing the LP coefficients, the corresponding LSP frequencies are quantized. After quantizing the LSP frequencies, if the properties of the LSP polynomials are preserved, the reconstructed LPC filter retains the minimum phase property [62]. The LSP frequencies are also called Line Spectral Frequencies (LSFs). The LSFs are quantized using an independent, non-uniform scalar quantization procedure. Scalar quantization may result in non-monotonicity of the LSFs and in that case, adjustments are performed in order to restore their monotonicity. The block diagram in Figure 3.1 shows the steps involved in the generation the LSP polynomials, computing the LSFs and quantizing them.

Construction of LSP polynomials and computation of their zeros [62] by applying Descartes' rule will be illustrated using MATLAB programs in the following sections. Correction of ill-conditioned cases that occur due to non-minimum phase LP polynomials and quantization of LSFs are also illustrated.

3.1 CONSTRUCTION OF LSP POLYNOMIALS

The LSP polynomials are computed from the linear predictor polynomial $A(z)$ as follows,

$$P(z) = A(z) + z^{-(m+1)}A(z^{-1}) , \qquad (3.1)$$
$$Q(z) = A(z) - z^{-(m+1)}A(z^{-1}) , \qquad (3.2)$$

where $P(z)$ is called the *symmetric polynomial* (palindromic) and $Q(z)$ is called the *anti-symmetric polynomial* (anti-palindromic). This is because of the reason that $P(z)$ has symmetric coefficients and $Q(z)$ has anti-symmetric coefficients. Evidently, $P(z)$ has a root at $(-1, 0)$ in the z-plane as

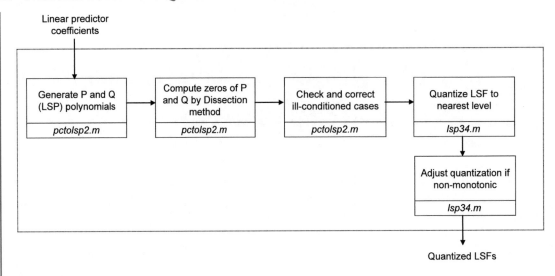

Figure 3.1: Computation of line spectral frequencies.

it has symmetric coefficients and $Q(z)$ has a root at $(0, 0)$ in the z-plane as it has anti-symmetric coefficients. To regenerate $A(z)$ from $P(z)$ and $Q(z)$, we use,

$$A(z) = 0.5[P(z) + Q(z)] . \tag{3.3}$$

In practice, only half the coefficients of the LSP polynomials, excluding the first element are stored. The reason is that the polynomials have symmetry and their first coefficient is equal to 1. In Program P3.1, in addition to storing a part of the coefficients as described above, all the coefficients of both $P(z)$ and $Q(z)$ are stored in the output variables pf and qf for illustration purposes.

Example 3.1 The linear predictor coefficients for the speech signal given in Figure 2.4 are shown in Figure 3.2. The resulting LSP polynomial coefficients are shown in Figure 3.3. Figure 3.3 shows that the coefficients of $P(z)$ are symmetric and those of $Q(z)$ are anti-symmetric.

As another example, a fourth-order LP polynomial is considered. Using Program P3.1, the coefficients of $P(z)$ and $Q(z)$ are generated. The coefficients of LP polynomial (a_4exp), the coefficients of the symmetric LSP polynomial (pf_4) and the coefficients of the anti-symmetric LSP polynomial (qf_4), are listed below.

$$a\_4exp = [1.0000; -1.5419; 1.3523; -0.7197; 0.3206] ,$$
$$pf\_4 = [1.0000; -1.2212; 0.6326; 0.6326; -1.2212; 1.0000] ,$$
$$qf\_4 = [1.0000; -1.8625; 2.0719; -2.0719; 1.8625; -1.0000] .$$

```
% P3.1   - generate_lsp.m

function [p,q,pf,qf]=generate_lsp(a,m)
% a      - LPC coefficients
% m      - LPC predictor order(here 10)
% p,q    - First half of coefficients of P(z)& Q(z),excluding 1
% pf,qf  - All the coefficients of P(z)& Q(z)
% MAXORD- Maximum order of P or Q(here 24)
MAXORD = 24;
MAXNO = m;


% INIT LOCAL VARIABLES
p = zeros( MAXORD, 1 );
q = zeros( MAXORD, 1 );
mp = m + 1;
mh = fix ( m/2 );
% GENERATE P AND Q POLYNOMIALS
p( 1:mh ) = a( 2:mh+1 ) + a( m+1:-1:m-mh+2 );
q( 1:mh ) = a( 2:mh+1 ) - a( m+1:-1:m-mh+2 );
p=[1;p(1:mh);flipud(p(1:mh));1];
q=[1;q(1:mh);-flipud(q(1:mh));-1];
% ALL THE COEFFICIENTS
pf=[1;p(1:mh);flipud(p(1:mh));1];
qf=[1;q(1:mh);-flipud(q(1:mh));-1];
```

Program P3.1: LSP polynomial construction.

3.2   COMPUTING THE ZEROS OF THE SYMMETRIC POLYNOMIAL

It is not possible to find the roots of the symmetric polynomial in closed form, but there are some properties of the polynomial that can be exploited to find its roots numerically in a tractable manner. The symmetric polynomial is given by,

$$P(z) = 1 + p_1 z^{-1} + p_2 z^{-2} + \cdots + p_2 z^{-(m-1)} + p_1 z^{-m} + z^{-(m+1)} , \qquad (3.4)$$

where p_i is the ith coefficient of $P(z)$. Assuming that the symmetric polynomial has roots lying on the unit circle, we evaluate the zeros of $P(z)$ by solving for ω in,

$$P(z)_{z=e^{j\omega}} = 2e^{-j\frac{m+1}{2}\omega} \left[\cos \frac{(m+1)\omega}{2} + p_1 \cos \frac{(m-1)\omega}{2} + \cdots + P_{(m+1)/2} \right] . \qquad (3.5)$$

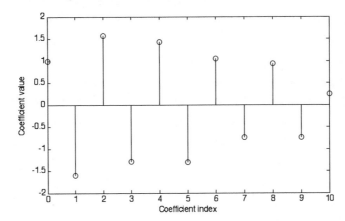

Figure 3.2: Linear predictor coefficients.

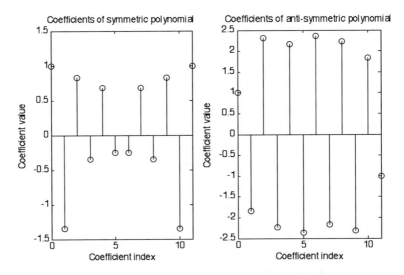

Figure 3.3: Symmetric and anti-symmetric polynomial coefficients.

Let us call the bracketed term of (3.5) as $P'(\omega)$. Since the zeros occur in complex conjugate pairs we compute only the zeros on the upper half of the unit circle. Also, the zero at the point $(-1, 0)$ in the unit circle is not computed. The MATLAB code to compute the zeros of the symmetric polynomial is given in Program P3.2.

   In order to compute the roots of the symmetric polynomial numerically using Descartes' rule, the following procedure is adopted. The upper half of the unit circle is divided into 128 intervals and each interval is examined for zeros of the polynomial. The fundamental assumption is that

```
% P3.2 - sym_zero.m

function [freq,sym_freq]=sym_zero(p,m)
% m    - Order of LPC polynomial
% freq- All LSP frequencies(to be computed)
% sym_freq - LSP frequencies of P(z)
% N    - Number of divisions of the upper half unit circle
% DEFINE CONSTANTS
EPS = 1.00e-6; % Tolerance of polynomial values for finding zero
N = 128;
NB = 15;
% INITIALIZE LOCAL VARIABLES
freq = zeros( m+1, 1 );
mp = m + 1;
mh = fix ( m/2 );
% COMPUTE P AT F=0
fl = 0.0;
pxl = 1.0 + sum( p(1:mh) );
% SEARCH FOR ZEROS OF P
nf = 1;
i = 1;
while i <= N
    mb = 0;
    % HIGHER FREQUENCY WHERE P IS EVALUATED
    fr = i * ( 0.5 / N );
    pxr = cos( mp * pi * fr );
    jc = mp - ( 2 * ( 1:mh )' );
    % ARGUMENT FOR COS
    ang = pi * fr * jc;
    % EVALUATION OF P AT HIGHER FREQUENCY
    pxr = pxr + sum( cos(ang) .* p(1:mh) );
    tpxr = pxr;
    tfr = fr;
    % COMPARE THE SIGNS OF POLYNOMIAL EVALUATIONS
    % IF THEY ARE OPPOSITE SIGNS, ZERO IN INTERVAL
    if ( pxl * pxr ) <= 0.00
        mb = mb + 1;
        % BISECTION OF THE FREQUENCY INTERVAL
        fm = fl + ( fr - fl ) / ( pxl - pxr ) * pxl;
        pxm = cos( mp * pi * fm );
        jc = mp - ( (1:mh)' * 2 );
        ang = pi * fm * jc;
        pxm = pxm + sum( cos(ang) .* p(1:mh) );
```

Program P3.2: Computation of zeros for symmetric LSP polynomial. (*Continues.*)

```
        % CHECK FOR SIGNS AGAIN AND CHANGE THE INTERVAL
        if ( pxm * pxl ) > 0.00
            pxl = pxm;
            fl = fm;
        else
            pxr = pxm;
            fr = fm;
        end
       % PROGRESSIVELY MAKE THE INTERVALS SMALLER AND REPEAT THE
       % PROCESS UNTIL ZERO IS FOUND
       while (abs(pxm) > EPS) & ( mb < 4 )
            mb = mb + 1;
            fm = fl + ( fr - fl ) / ( pxl - pxr ) * pxl;
            pxm = cos( mp * pi * fm );
            jc = mp - ( (1:mh)' * 2 );
            ang = pi * fm * jc;
            pxm = pxm + sum( cos(ang) .* p(1:mh) );
            if ( pxm * pxl ) > 0.00
                pxl = pxm;
                fl = fm;
            else
                pxr = pxm;
                fr = fm;
            end
       % IF THE PROCESS FAILS USE DEFAULT LSPS
        if ( (pxl-pxr) * pxl ) == 0
            freq( 1:m ) = ( 1:m ) * 0.04545;
            fprintf( 'pctolsp2: default lsps used, avoiding /0\n');
            return
        end
        % FIND THE ZERO BY BISECTING THE INTERVAL
        freq(nf) = fl + (fr-fl) / (pxl-pxr) * pxl;
        nf = nf + 2;
        if nf > m-1
            break
        end
    end
    pxl = tpxr;
    fl = tfr;
    i = i + 1;
end
% LSP Frequencies of P(z) alone
sym_freq=freq(1:2:m);
```

Program P3.2: (*Continued.*) Computation of zeros for symmetric LSP polynomial.

the interval is so small that it contains at most a single zero. The polynomial $P'(\omega)$ is evaluated at the frequencies fr and fl which are the end frequencies of the interval considered. The values of the polynomial $P'(\omega)$ in the two frequencies are stored in the program variables pxr and pxl, respectively.

Initially the value of $P'(\omega)$ at frequency 0 is the sum of its coefficients. Then $P'(\omega)$ is evaluated at the radian frequency ω, where ω is represented as $\pi i/128$. In general, the program variable jc contains the argument of the cosine function excluding 0.5ω, and the program variable ang is the complete argument for the cosine function.

Upon computation of pxr and pxl, their signs are compared. If there is a sign change, then by Descartes' rule, there are an odd number of zeros. However, as our interval is small enough there can be no more than a single zero. If there is no sign change, there are no zeros and therefore the program proceeds to the next interval. If there is a zero, the next step is the computation of mid-frequency fm using the bisection method. The value of pxm is subsequently calculated by using fm. Then again, the signs of pxm and pxl are compared. If there is a sign change, then there is a zero between fm and fl. Therefore, the interval is changed accordingly. This is repeated until the value of pxm gets close to zero and then the frequency interval is bisected to find the zero. After finding a zero, we move on to the next interval to find the next zero. This process is repeated until all the zeros on the upper half of the unit circle are determined. Every time a zero is calculated, the number of frequencies computed (nf) is incremented by two because each zero on the upper half of the unit circle is a complex conjugate of another in the lower half of the unit circle. Hence, finding one zero is actually equivalent to finding two LSFs.

Example 3.2   The roots of the symmetric polynomial whose coefficients are shown in Figure 3.3 are given in Figure 3.4. As it can be seen, the roots lie on the unit circle and appear as complex conjugate pairs.

3.3   COMPUTING THE ZEROS OF THE ANTI-SYMMETRIC POLYNOMIAL

Finding the root of the anti-symmetric polynomial is similar to finding the roots of the symmetric polynomial. However, it is simpler because of the property that the roots of the symmetric and anti-symmetric polynomial are interlaced. The anti-symmetric polynomial is given by,

$$Q(z) = 1 + q_1 z^{-1} + q_2 z^{-2} \ldots - q_2 z^{-(m-1)} - q_1 z^{-m} - z^{-(m+1)} , \qquad (3.6)$$

where q_i's are the coefficients of $Q(z)$. Since we need to evaluate the zeros of $Q(z)$ only on the unit circle, we can write $Q(z)$ as,

$$Q(z)_{z=e^{j\omega}} = 2je^{-j\frac{m+1}{2}\omega} \left[\sin\frac{(m+1)\omega}{2} + q_1 \sin\frac{(m-1)\omega}{2} + \ldots + q_{(m+1)/2} \right] , \qquad (3.7)$$

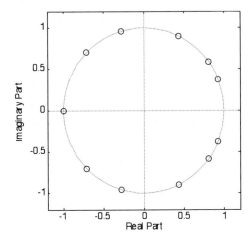

Figure 3.4: Roots of the symmetric LSP polynomial.

where we call the bracketed term as $Q'(\omega)$. The zero at the point $(0, 0)$ in the unit circle is not computed. The procedure to compute the roots of $Q'(\omega)$ and hence that of $Q(z)$ is given in Program P3.3.

The procedure for computing the zeros of $Q(z)$ is very similar to the procedure for computing the zeros of $P(z)$, except for a few differences. Firstly, since we know that the zeros of $P(z)$ and $Q(z)$ are interlaced, the start and end points of the interval are taken as the consecutive zeros of symmetric LSP polynomial. Secondly, instead of using the cosine function, we use sine function because of the difference between (3.5) and (3.7).

Example 3.3   The roots of the anti-symmetric polynomial whose coefficients are shown in Figure 3.3 are given in Figure 3.5. Again, all the roots lie on the unit circle and appear as complex conjugate pairs.

Another example is to consider a fourth-order LP polynomial and compute the LSP frequencies. Considering the same LP polynomial in Example 3.1, we get the symmetric LSFs (`sym_freq_4`) from Program P3.2 and anti-symmetric LSFs (`asym_freq_4`) from Program P3.3 as,

$$\texttt{sym\_freq\_4} = [0.0842; 0.2102],$$
$$\texttt{asym\_freq\_4} = [0.1082; 0.3822],$$
$$\texttt{freq\_4} = [0.0682; 0.1082; 0.2033; 0.3822],$$

```
% P3.3 - antisym_zero.m

function [freq,asym_freq]=antisym_zero(q,m,freq)
% m    - Order of LPC polynomial
% freq- LSP frequencies(to be computed)
% N    - Number of divisions of the upper half unit circle
% EPS - Tolerance of polynomial values for finding zeros
% DEFINE CONSTANTS
EPS = 1.00e-6;
N = 128;
NB = 15;

% INITIALIZE LOCAL VARIABLES
mp = m + 1;
mh = fix ( m/2 );

% THE LAST FREQUENCY IS 0.5
freq(m+1) = 0.5;
% FIRST SYMMETRIC LSP FREQUENCY IS THE STARTING POINT
fl = freq(1);
qxl = sin( pi * mp * fl );
jc = mp - ( (1:mh)' * 2 );
% ARGUMENT FOR SIN
ang = pi * fl * jc;
qxl = qxl + sum( sin(ang) .* q(1:mh) );

i = 2;
while i < mp
    mb = 0;
    % USE ONLY SYMMETRIC LSP FREQUENCIES AS END POINTS(ZEROS INTERLACED)
    fr = freq(i+1);
    qxr = sin( mp * pi * fr );
    jc = mp - ( (1:mh)' * 2 );
    ang = pi * fr * jc;
    qxr = qxr + sum( sin(ang) .* q(1:mh) );
    tqxl = qxl;
    tfr = fr;
    tqxr = qxr;
    mb = mb + 1;
    fm = ( fl + fr ) * 0.5;
    qxm = sin( mp * pi * fm );
    jc = mp - ( (1:mh)' * 2 );
    ang = pi * fm * jc;
```

Program P3.3: Computation of zeros for anti-symmetric LSP polynomial. (*Continues.*)

```
    qxm = qxm + sum( sin(ang)  .* q(1:mh) );
    % CHECK FOR SIGN CHANGE
    if ( qxm * qxl ) > 0.00
        qxl = qxm;
        fl = fm;
    else
        qxr = qxm;
        fr = fm;
    end
    % Progressively make the intervals smaller and repeat the
    % process until zero is found
    while ( abs(qxm) > EPS*tqxl ) & ( mb < NB )
        mb = mb + 1;
        fm = ( fl + fr ) * 0.5;
        qxm = sin( mp * pi * fm );
        jc = mp - ( (1:mh)' * 2 );
        ang = pi * fm * jc;
        qxm = qxm + sum( sin(ang)  .* q(1:mh) );
        if ( qxm * qxl ) > 0.00
            qxl = qxm;
            fl = fm;
        else
            qxr = qxm;
            fr = fm;
        end
    end
    % If the process fails use previous LSPs
    if ( ( qxl - qxr ) * qxl ) == 0
        freq( 1:m ) = ( 1:m ) * 0.04545;
        fprintf( 'pctolsp2: default lsps used, avoiding /0\n');
        return
    end
    % Find the zero by bisecting the interval
    freq(i) = fl + ( fr - fl ) / ( qxl - qxr ) * qxl;
    qxl = tqxr;
    fl = tfr;
    i = i + 2;
end
% LSP frequencies excluding the last one(0.5)
freq=freq(1:m);
% LSP Frequencies of Q(z) alone
asym freq=freq(2:2:m);
```

Program P3.3: (*Continued.*) Computation of zeros for anti-symmetric LSP polynomial.

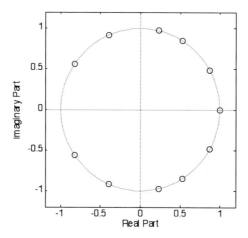

Figure 3.5: Roots of the anti-symmetric LSP polynomial.

where the array `freq_4` stores all the LSFs. Note that the LSFs are normalized with respect to the sampling frequency, and hence they range between 0 and 0.5. The relation between an LSF and the corresponding zero of the LSP polynomial is given as,

$$\hat{z} = \cos(2\pi f) + i \sin(2\pi f) , \qquad (3.8)$$

where f is the LSF and $\hat{z}$ is the corresponding zero of the LSP polynomial. The LSFs 0 and 0.5 are not included as it is evident that the zero of $P(z)$ at the point $(0, 0)$ in the unit circle corresponds to a normalized frequency of 0 and the zero of $Q(z)$ at the point $(-1, 0)$ in the unit circle corresponds to a normalized frequency of 0.5. The LSFs for the LP polynomial given in Example 3.1 are given in Figure 3.6, where stars indicate the LSFs from $P(z)$ and circles indicate the LSFs from $Q(z)$.

So far, we have examined LP polynomials that are minimum phase to generate LSP frequencies. We study here the effect of LP polynomials with non-minimum phase by considering the following coefficients of $A(z)$,

$$\texttt{a\_nm} = [1.0000; -1.5898; 0.4187; 0.3947; 0.4470] .$$

The pole-zero plot of $1/A(z)$ is given in Figure 3.7, where we can see the poles lying outside the unit circle making $A(z)$ a non-minimum phase polynomial. Figure 3.8 shows the LSFs of the symmetric LSP polynomial (shown as stars) and the LSFs of anti-symmetric LSP polynomial (shown as circles). Note that the frequencies are not obtained using Programs P3.2 and P3.3, but rather they are calculated directly using MATLAB. We can see that all the roots of $Q(z)$ do not lie on the unit circle and hence do not satisfy the assumption that all the roots of the LSP polynomials lie on the unit circle. Therefore, the LSFs obtained from the Program P3.2 and P3.3 will not be

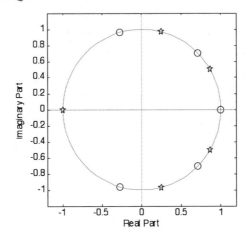

Figure 3.6: Roots of the LSP polynomials.

accurate because of the assumption of minimum phase LP polynomial, $A(z)$. The attempt to detect and correct this case of ill-condition is described in the next section.

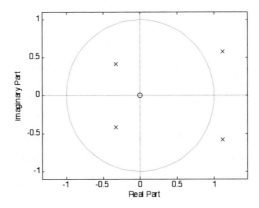

Figure 3.7: Pole-zero plot of the non-minimum phase filter.

3.4   TESTING ILL-CONDITIONED CASES

Ill-conditioned cases in finding the roots of the LSP polynomials occur if the LP polynomial used to generate them is non-minimum phase. In a well-conditioned case, the LP polynomial is minimum phase and our assumption that the zeros of LSP polynomials lie on the unit circle is valid. If the zeros of the symmetric and anti-symmetric LSP polynomials do not alternate on the unit circle, the

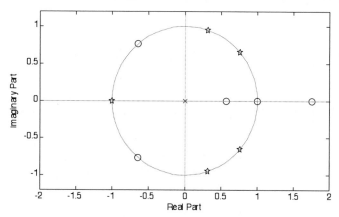

Figure 3.8: Roots of the LSP polynomials for the non-minimum phase filter.

LP polynomial may not be minimum phase. Therefore, the Programs P3.2 and P3.3 do not output monotonic LSFs.

Program P3.4 tests ill-conditioned cases that occur with LSPs and tries to correct them. If more than one LSF is 0 or 0.5, then the program variable `lspflag` is set to 1, indicating that a case of ill-condition is encountered. If the LSFs are non-monotonic, an attempt to correct them is made. If that attempt fails, the LSFs of the previous frame are used.

Example 3.4   Consider the non-minimum phase LP polynomial in Example 3.3, and using Programs P3.1, P3.2, and P3.3, we obtain the LSFs as,

$$\texttt{freq} = [0.1139; 0.0064; 0.1987; 0.3619].$$

We can see that the LSFs are non-monotonic, and using Program P3.4, we can detect and correct the ill-conditioned case. After using Program P3.4, the LSFs re-ordered for monotonicity are obtained as,

$$\texttt{freq} = [0.0064; 0.1139; 0.1987; 0.3619].$$

3.5   QUANTIZING THE LINE SPECTRAL FREQUENCIES

The LSFs will be quantized using the quantization matrix given in Table 3.1. Six frequencies (LSF 1 and LSFs 5 to 9) will be quantized using 3 bits and the remaining four (LSFs 2 to 5) will be quantized

Table 3.1: LSF quantization matrix.

Frequency Index	Quantization Levels														
100	170	225	250	280	340	420	500	0	0	0	0	0	0	0	0
210	235	265	295	325	360	400	440	480	520	560	610	670	740	810	880
420	460	500	540	585	640	705	775	850	950	1050	1150	1250	1350	1450	1550
620	660	720	795	880	970	1080	1170	1270	1370	1470	1570	1670	1770	1870	1970
1000	1050	1130	1210	1285	1350	1430	1510	1590	1670	1750	1850	1950	2050	2150	2250
1470	1570	1690	1830	2000	2200	2400	2600	0	0	0	0	0	0	0	0
1800	1880	1960	2100	2300	2480	2700	2900	0	0	0	0	0	0	0	0
2225	2400	2525	2650	2800	2950	3150	3350	0	0	0	0	0	0	0	0
2760	2880	3000	3100	3200	3310	3430	3550	0	0	0	0	0	0	0	0
3190	3270	3350	3420	3490	3590	3710	3830	0	0	0	0	0	0	0	0

```
% P3.4 - lsp_condition.m
function [ freq ] = lsp_condition( freq,lastfreq,m )
% freq - LSP Frequencies of current frame
% lastfreq - LSP frequencies of last frame
% m - Order of LP polynomial

% TEST FOR ILL-CONDITIONED CASES AND TAKE CORRECTIVE ACTION IF REQUIRED
freq = freq(1:m);
lspflag = 0;
if ( any(freq==0.00) | any(freq==0.5) )
    lspflag = 1;
end
% REORDER LSPs IF NON-MONOTONIC
for i = 2:m
    if freq(i) < freq(i-1)
        lspflag = 1;
        fprintf( 'pctolsp2: non-monotonic lsps\n' );
        tempfreq = freq(i);
        freq(i) = freq(i-1);
        freq(i-1) = tempfreq;
    end
end
% IF NON-MONOTONIC AFTER 1ST PASS, RESET TO VALUES FROM PREVIOUS FRAME
for i = 2:m
    if freq(i) < freq(i-1)
        fprintf( 'pctolsp2: Reset to previous lsp values\n' );
        freq(1:m) = lastfreq;
        break;
    end
end
```

Program P3.4: Testing cases of ill-condition.

using 4 bits resulting in a total of 34 bits per frame. The LSFs are quantized such that the absolute distance of each unquantized frequency with the nearest quantization level is minimal.

The first LSF is quantized using the quantization levels given in the first row of the matrix given in Table 3.1, the second LSF using the second row and so on. Note that in some cases though the second LSF is higher than the first, it may be quantized to a lower value. This is because the quantization tables overlap, i.e., the non-zero values in each of the columns of Table 3.1 are not always in increasing order. Nevertheless, quantization is performed using the quantization table, and the issue of quantized LSFs becoming possibly non-monotonic is handled later. Also, it can be seen from Table 3.1, that the number of levels used for quantizing the second, third, fourth and fifth LSFs are 16, thereby needing 4 bits; whereas for the other LSFs, there are 9 quantization levels, hence needing 3 bits each.

In Program P3.5, the program variable `bits` is initialized as,

$$\text{bits} = [3; 4; 4; 4; 4; 3; 3; 3; 3; 3]\,.$$

The quantization matrix given in Table 3.1 is stored in the program variable lspQ. FSCALE is the sampling frequency, which is used to scale the LSFs to the appropriate range. For each LSF, the number of bits used for quantization determines the number of levels. Quantization of each LSF is performed such that the distance between the unquantized LSF and the corresponding quantization level in the table is minimized. The index of the quantized LSF is stored in the program variable findex.

```
% P3.5  - lsp_quant.m
function [ findex ] = lsp_quant( freq, no, bits, lspQ )
% freq - Unquantized LSFs
% no - Number of LSFs (10 here)
% bits - Array of bits to be allocated for each LSF
% findex - Vector of indices to quantized LSFs, references lspQ
% lspQ - Quantization table matrix

% DEFINE CONSTANTS
FSCALE = 8000.00;
MAXNO = 10;
% INIT RETURN VECTOR
findex = zeros( MAXNO, 1 );
% INIT LOCAL VARIABLES
freq = FSCALE * freq;
levels = ( 2 .^ bits ) - 1;
% QUANTIZE ALL LSP FREQUENCIES AND FORCE MONOTONICITY
for i = 1:no
    % QUANTIZE TO NEAREST OUTPUT LEVEL
    dist = abs( freq(i) - lspQ(i,1:levels(i)+1) );
    [ low, findex(i) ] = min( dist );
end
```

Program P3.5: Quantizing the LSFs.

Example 3.5   The LSFs are quantized using Program P3.5 to obtain indices of quantized LSFs (findex) in the quantization table. Table 3.2 gives the input unquantized LSFs, unquantized frequencies, output indices (findex) and the quantized frequencies corresponding to those indices. The spectral plots of the LP synthesis filters corresponding to the unquantized and quantized LSFs are given in Figure 3.9. The corresponding pole-zero plots are given in Figure 3.10. Note that the LSFs must be converted to LP coefficients using Program P4.1 in order to visualize the spectra and pole-zero plots.

Table 3.2:	Example results for	Program P3.5.	
Unquantized LSF	Unquantized Frequency (Hz)	Index to Quantization Table	Quantized Frequency (Hz)
0.0617	493.3	8	500
0.0802	641.3	13	670
0.0999	799.1	8	775
0.1612	1289.7	9	1270
0.1785	1428.0	7	1430
0.2128	1702.2	3	1690
0.2957	2365.9	5	2300
0.3141	2513.0	3	2525
0.3761	3009.1	3	3000
0.4049	3239.4	2	3270

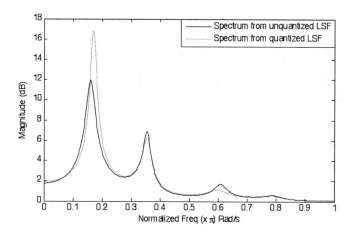

Figure 3.9: Spectra from unquantized and quantized LSFs.

3.6   ADJUSTING QUANTIZATION TO PRESERVE MONOTONICITY

Quantization of LSFs may lead to non-monotonicity because of the overlapped quantization tables. But, it is important to preserve the monotonicity of LSFs because if they become non-monotonic, the zeros of the two LSP polynomials will no longer be interlaced. This will lead to the loss of minimum phase property of the LP polynomial. Therefore, the quantized LSFs are checked for monotonicity, and if there is a case where the previous quantized LSF has a higher value than the

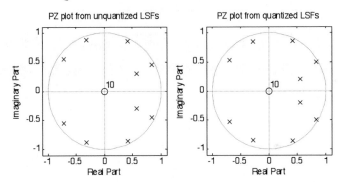

Figure 3.10: Pole-zero plots from unquantized and quantized LSFs.

current LSF, then the problem is corrected by either quantizing the current LSF to the next higher level or quantizing the previous LSF to the next lower level.

Unquantized LSFs	Index to Quantization Table	Non-monotonic Quantized LSFs	Corrected Index to Quantization Table	Monotonic Quantized LSFs
0.0501	7	0.0525	7	0.0525
0.0512	7	0.0500	8	0.055
0.0942	8	0.0969	8	0.0969
0.1209	6	0.1212	6	0.1212
0.1387	3	0.1413	3	0.1413
0.2335	4	0.2288	4	0.2288
0.3103	6	0.3100	6	0.31
0.3214	3	0.3156	3	0.3156
0.4235	7	0.4288	7	0.4288
0.433	5	0.4363	5	0.4363

Table 3.3: Example results for Program P3.6.

Program P3.6 takes unquantized LSFs (`freq`) and quantized LSF indices (`findex`) as the first two inputs. For each LSF, the program checks whether the current quantized LSF is less than or equal to the previous quantized LSF. If this is true, then the corresponding upward and downward errors are calculated. Upward error is the error resulting if the current LSF were quantized to a next higher value, with the previous LSF remaining at the same quantized level. Downward error is the error resulting if the previous LSF were quantized to a next lower value, with the current LSF remaining at the same quantized level. If the upward error is lower than the downward error, then the current LSF is quantized to a next higher level. Else, the previous LSF is quantized to the next

```
% P3.6  - lsp_adjust_quant.m
function [ findex ] = lsp_adjust_quant( freq,findex,no,lspQ,bits )
% freq - Unquantized LSFs
% findex - Vector of indices to quantized LSFs, references lspQ
%          (May index non-monotonic LSFs on input, but will index to
%           monotonic LSFs on output)
% lspQ - Quantization table matrix
% no - Number of LSFs (10 here)
% bits - Array of bits to be allocated for each LSF

% DEFINE CONSTANTS
FSCALE = 8000.00;
MAXNO = 10;
% INIT LOCAL VARIABLES
freq = FSCALE * freq;
levels = ( 2 .^ bits ) - 1;
% QUANTIZE ALL LSP FREQUENCIES AND FORCE MONOTONICITY
for i = 1:no
    % ADJUST QUANTIZATION IF NON-MONOTONICALLY QUANTIZED AND
    % FIND ADJUSTMENT FOR MINIMUM QUANTIZATION ERROR
    if i > 1
        if lspQ( i,findex(i) ) <= lspQ( i-1,findex(i-1) )
            errorup = abs(freq(i)- ...
                lspQ(i,min(findex(i)+1,levels(i))))+ ...
                abs( freq(i-1) - lspQ(i-1, findex(i-1)) );
            errordn = abs( freq(i)    - lspQ(i,findex(i)) ) + ...
                abs( freq(i-1) - lspQ(i-1,max(findex(i-1)-1,0)) );
            % ADJUST INDEX FOR MINIMUM ERROR(AND PRESERVE MONOTONICITY)
            if errorup < errordn
                findex(i) = min( findex(i)+1, levels(i) );
                while ( lspQ(i,findex(i)) < lspQ(i-1,findex(i-1)) )
                    findex(i) = min( findex(i)+1, levels(i) );
                end
            elseif i == 1
                findex(i-1) = max( findex(i-1)-1, 0 );
            elseif lspQ(i-1,max(findex(i-1)-1,0))>lspQ(i-2,findex(i-2))
                findex(i-1) = max( findex(i-1)-1, 0 );
            else
                findex(i) = min( findex(i)+1, levels(i) );
                while lspQ( i, findex(i) ) < lspQ( i-1, findex(i-1) )
                    findex(i) = min( findex(i)+1, levels(i) );
                end
            end
        end
    end
end
```

Program P3.6: Adjusting quantization to preserve monotonicity.

lower level. If this operation introduces non-monotonicity, then the current LSP is quantized to the next higher value, regardless of the higher upward error.

Example 3.6   In this example, we will consider a case where the quantized LSFs are non-monotonic as observed in Table 3.3. If we provide the indices of the non-monotonically quantized LSFs (findex) shown in the second column of Table 3.3 as an input to Program P3.6, we get the indices shown in fourth column as the output. The indices to the quantization table output from the program correspond to monotonic quantized LSFs that are given in the last column of Table 3.3. The pole-zero plots of the LP synthesis filters corresponding to unquantized LSFs, quantized LSFs and quantized LSFs corrected to be monotonic are given in Figure 3.11. The pole that crosses the unit circle is zoomed in and it is clear that the non-monotonicity of LSFs that occurs due to quantization results in the pole crossing to the outside of the unit circle. When the LSFs are corrected to be monotonic, the pole comes inside the unit circle again, making the system minimum phase.

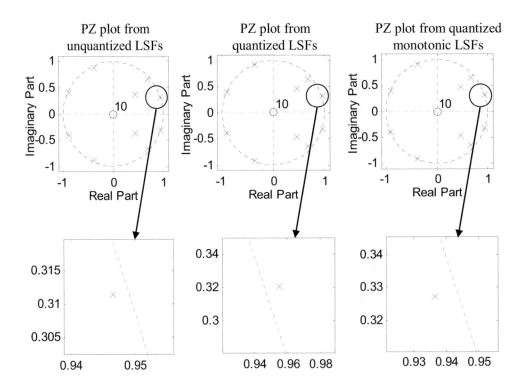

Figure 3.11: Pole-zero plots from unquantized, quantized and monotonically quantized LSFs.

3.7   SUMMARY

The LSP polynomials, $P(z)$ and $Q(z)$ are generated from the LP coefficients. $P(z)$ is a polynomial with even-symmetric coefficients and $Q(z)$ is a polynomial with odd symmetric coefficients. The zeros of the polynomials, the LSFs, are computed using Descartes' method. LSFs can be efficiently computed using Chebyshev polynomials [63], and details on the spectral error characteristics and perceptual sensitivity of LSFs are given in a paper by Kang and Fransen [64]. Developments in enhancing the coding of the LSFs include using vector quantization [65] and discrete cosine transform [66] to encode the LSFs. In third-generation standards, vector quantization of the LSFs [67, 68] is common practice. Note that ill-conditioned cases are indicated by the non-monotonicity of the LSFs. They could arise either due to a non-minimum phase LP polynomial or quantization of LSFs. Immittance Spectral Pairs (ISPs) are another form of spectral representation for the LP filter [69] similar to LSPs, and they are considered to represent the *immitance* of the vocal tract. Some of the most modern speech coding standards such as the AMR-WB codec and the source-controlled VMR-WB codec use ISPs as parameters for quantization and transmission [5, 42].

CHAPTER 4

Spectral Distortion

Spectral distortion in the FS-1016 encoder occurs due to the quantization of LSFs. The quantized and unquantized LSFs are converted to their respective LP coefficients. The LP coefficients are then converted to the RCs using inverse Levinson-Durbin recursion. The RCs are finally converted back to autocorrelation lags and the distances between the autocorrelation lags are computed. There are many distance measures that could be used to characterize the distortion due to the quantization of the LSFs. The following distance measures are considered in this chapter: (1) the cepstral distance measure, (2) the likelihood ratio and (3) the cosh measure. The block diagram in Figure 4.1 shows the steps involved in the computation of the spectral distortion measures.

Computation of the RCs from LP coefficients using inverse Levinson-Durbin recursion has been already explained in Section 2.5 (Program P2.5). Conversion of LSPs to LP coefficients, conversion of RCs to autocorrelation lags and calculation of distances between autocorrelation lags will be illustrated in the following sections using MATLAB programs.

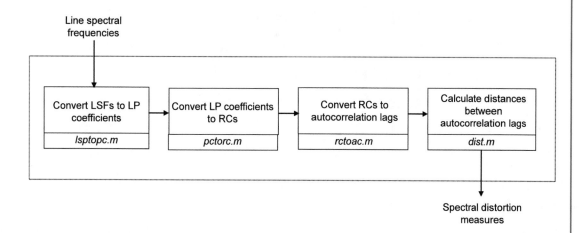

Figure 4.1: Spectral distortion computation.

4.1   CONVERSION OF LSP TO DIRECT-FORM COEFFICIENTS

Conversion of LSP frequencies to LP coefficients is much less complex than finding the LSP frequencies from the LP polynomial. Each LSP frequency ω_i represents a polynomial of the form $1 - 2\cos\omega_i z^{-1} + z^{-2}$ [63]. Upon multiplying the second-order factors, we get the symmetric and anti-symmetric polynomials $P(z)$ and $Q(z)$. Then the LP polynomial can be reconstructed using (3.3). But here we adopt a different approach by realizing that each second-order factor of the polynomial $P(z)$ and $Q(z)$ can be considered as a second-order filter, of the type shown in Figure 4.2. The total filter polynomial is a cascade of many such second-order stages. Because the filter is Finite Impulse Response (FIR), the impulse response samples are the coefficients of the filter. Therefore, the coefficients of the LP polynomials are found by computing the impulse response of the cascade of the second-order stages.

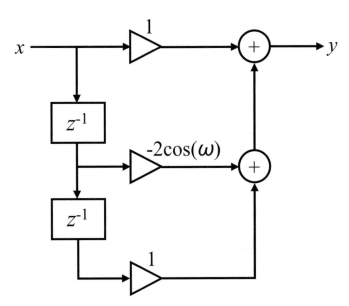

Figure 4.2: Second-order section of $P(z)$ and $Q(z)$.

In Program P4.1, p and q are the arrays that contain the cosine of the LSP frequencies of the polynomials $P(z)$ and $Q(z)$. The variables a and b are the arrays that contain the outputs of the current stage and, consequently, the inputs for the next stage. The variables a_1 and b_1 are the first level memory elements that correspond to first-order delays and a_2 and b_2 are the second level memory elements that correspond to second-order delays. The outer for loop cycles through the samples and the inner for loop cycles through the stages of the cascaded filter for each input sample. The latter calculates the output samples for each stage and populates the memory elements. Each LP polynomial coefficient, except the first coefficient, which is always one, is computed as the average of the coefficients of the two filters as per (3.3). Finally, the LP coefficients are stored in the array pc.

Example 4.1   The Program P4.1 takes the LSP frequencies as the input and generates the LP coefficients as the output. For the case of a tenth-order LP polynomial, quantized and unquantized LSP frequencies (f and fq) supplied as input to the program and the corresponding LP coefficients (pc and pcq) output from the program are given below. The spectra and pole-zero plots of the LP synthesis filters corresponding to unquantized and quantized LSFs are given in Figures 3.9 and 3.10, respectively. The LP coefficients (pc and pcq) are converted to RCs using Program P2.5, and the corresponding RCs (rc and rcq) are given below.

$$f = [0.0617; 0.0802; 0.0999; 0.1612; 0.1785; 0.2128; 0.2957; 0.3141; 0.3761; 0.4049],$$
$$pc = [1.0000; -1.5921; 1.5702; -1.2870; 1.4251; -1.3094; 1.0557; -0.7447; 0.9385; -0.7386; 0.2440],$$
$$fq = [0.0625; 0.0838; 0.0969; 0.1588; 0.1787; 0.2112; 0.2875; 0.3156; 0.3750; 0.4088],$$
$$pcq = [1.0000; -1.6440; 1.6310; -1.2326; 1.2500; -1.1169; 0.9365; -0.6354; 0.7458; -0.5828; 0.1708].$$

The LP coefficients (pc and pcq) are also converted to RCs using Program P2.5 and the corresponding RCs (rc and rcq) are also given below. Note that the absolute value of the elements of the arrays rc and rcq are less than one, which implies that the LP synthesis filter is stable.

$$rc = [0.7562; -0.6992; 0.2369; -0.3205; 0.0473; -0.5316; -0.1396; -0.0367; 0.3723; -0.2440],$$
$$rcq = [0.7817; -0.7778; 0.2507; -0.3701; -0.0816; -0.4893; -0.0247; 0.0151; 0.3111; -0.1708].$$

4.2   COMPUTATION OF AUTOCORRELATION LAGS FROM REFLECTION COEFFICIENTS

To calculate the Autocorrelation (AC) lags from RCs [50], we first obtain the LP polynomials of increasing orders from the reflection coefficients using forward and backward predictor recurrence relations that are given in (4.1) and (4.2), respectively,

$$a_{i+1}(j) = a_i(j) + k_{i+1}a_i(i + 1 - j) \tag{4.1}$$
$$a_{i+1}(i + 1 - j) = a_i(i + 1 - j) + k_{i+1}a_i(j), \tag{4.2}$$

where, $a_i(j)$ is the jth predictor coefficient of order i and k_i is the reflection coefficient. In Program P4.2, initially the autocorrelation array r is stored with the reflection coefficients rc. Then

```
% P4.1 - lsptopc.m
function pc = lsptopc( f, no )
%    f            -        LSP frequencies
%    no           -        LPC filter order
%    pc           -        LPC predictor coefficients

% INITIALIZE LOCAL VARIABLES
noh = no / 2;
freq = f;
a = zeros( noh+1, 1 );
a1 = a;
a2 = a;
b = a;
b1 = a;
b2 = a;
pc = zeros( no, 1 );
% INITIALIZE LSP FILTER PARAMETERS
p = -2 * cos( 2*pi*freq((1:2:no-1)') );
q = -2 * cos( 2*pi*freq((2:2:no)') );
% COMPUTE IMPULSE RESPONSE OF ANALYSIS FILTER
xf = 0.00;
for k = 1:no+1
    xx = 0.00;
    if k == 1
        xx = 1.00;
    end
    a(1) = xx + xf;
    b(1) = xx - xf;
    xf = xx;
```

Program P4.1: Conversion of LSP frequencies to LP coefficients. (*Continues.*)

```
for i = 1:noh
    a(i+1) = a(i) + ( p(i) * a1(i) ) + a2(i);
    b(i+1) = b(i) + ( q(i) * b1(i) ) + b2(i);
    a2(i) = a1(i);
    a1(i) = a(i);
    b2(i) = b1(i);
    b1(i) = b(i);
end
if k ~= 1
    pc(k-1) = -0.5 * ( a(noh+1) + b(noh+1) );
end
end
% CONVERT TO PREDICTOR COEFFICIENT ARRAY CONFIGURATION
pc(2:no+1) = -pc(1:no);
pc(1) = 1.0;
```

Program P 4.1: (*Continued.*) Conversion of LSP frequencies to LP coefficients.

a temporary array t is created to store the intermediate LP polynomial coefficients of increasing orders. If we have a predictor polynomial of order m, then we know that,

$$a_m(m) = k_m .$$ (4.3)

In Program P4.2, we negate the reflection coefficient and assign it to the intermediate LP polynomial coefficient because we want to follow the sign convention where the first reflection coefficient equals the normalized first-order autocorrelation lag. The LP polynomial coefficients are computed for increasing order of the intermediate predictor polynomials from 2 to m, where m is the order of the actual predictor polynomial. The order of the intermediate LP polynomial is indicated by the variable i in the program. For each intermediate predictor polynomial, the first half of the polynomial coefficients are computed using forward predictor recursion given in (4.1) and the second half of the polynomial coefficients are computed using reverse predictor recursion given in (4.2). The last coefficient of any intermediate polynomial is equal to the negative of reflection coefficient. This follows from (4.3) and the negative sign is because of our sign convention.

Then, the autocorrelation lag of order $i + 1$ is computed by,

$$r(i + 1) = \sum_{k=1}^{i+1} a_{i+1}(k)r(i + 1 - k) ,$$ (4.4)

where, $r(i)$ is the autocorrelation at lag i. In Program P4.2, all the coefficients of the intermediate LP polynomial except the first coefficient are negatives of their actual values because of our sign

```
% P4.2   - rctoac.m
function r = rctoac( rc, m )
%    rc          -          Reflection coefficients
%    m           -          Predictor order
%    r           -          Normalized autocorrelation lags


% INITIALIZE LOCAL VARIABLES
r = zeros( m+1, 1 );
t = r;
r(1) = 1.0;
r(2:m+1) = rc;


% COMPUTE PREDICTOR POLYNOMIAL OF DIFFERENT DEGREE AND STORE IN T
% COMPUTE AUTOCORRELATION AND STORE IN R
t(1) = 1.0;
t(2) = -r(2);
if m > 1
    for i = 2:m
        j = fix(i/2);
        tj = t( 2:j+1 ) - ( r( i+1 ) * t( i:-1:i-j+1 ) );
        tkj = t( i:-1:i-j+1 ) - ( r( i+1 ) * t( 2:j+1 ) );
        t( 2:j+1 ) = tj;
        t( i:-1:i-j+1 ) = tkj;
        t( i+1 ) = -r( i+1 );
        r( i+1 ) = r( i+1 ) - sum( t( 2:i ) .* r( i:-1:2 ) );
    end
end
```

Program P4.2: Computation of autocorrelation lags from reflection coefficients.

convention. Therefore, when (4.4) is implemented in the program, the final sum is negated to get the autocorrelation lag with proper sign.

Example 4.2   The autocorrelation lags that are output when the RCs in Example 4.1 are input to Program P4.2 are given below. The array ac contains the autocorrelation lags corresponding to rc and acq contains the autocorrelation lags corresponding to rcq. We will be using these autocorrelation lags to calculate various distance measures that quantify the amount of distortion that has occurred due to quantization.

ac = [1; 0.7562; 0.2725; −0.1267; −0.3443; −0.4001; −0.4154; −0.4701; −0.4573; −0.2241; 0.1523] ,
acq = [1; 0.7817; 0.3086; −0.1406; −0.4277; −0.5416; −0.5576; −0.5160; −0.3639; −0.0506; 0.3324] .

4.3   CALCULATION OF DISTANCE MEASURES

The distance measures used to measure the distortion due to quantization of LSP frequencies are the likelihood ratio, the cosh measure and the cepstral distance [70]. Nine different distances are stored in an array for each frame. In Program P4.3 (`dist.m`), the array of distance measures is indicated by dm.

Considering the log likelihood distance measure, let us define $A(z)$ as the LP polynomial that corresponds to the unquantized LSFs. The LP polynomial that would correspond to quantized LSFs is denoted by $A'(z)$. The minimum residual energy α would be obtained if the speech samples of the current frame, say $\{x(n)\}$, used to generate the LP polynomial $A(z)$ is passed through the same filter. The residual energy δ, is obtained when $\{x(n)\}$ is passed through the filter $A'(z)$. Consequently, if we assume that some sequence of samples $\{x'(n)\}$ was used to generate $A'(z)$, then α' would be the minimum residual energy found when $\{x'(n)\}$ was passed through $A'(z)$. The residual energy δ' is obtained when $\{x'(n)\}$ is passed through $A(z)$. The ratios δ'/α' and δ/α are defined as the likelihood ratios and indicate the differences between the LPC spectra before and after quantization.

The program `cfind.m` is used by `dist.m` to compute the cepstral coefficients, the filter autocorrelation lags, the residual energy, the LP filter parameters and the reflection coefficients, using the autocorrelation sequence. The method of finding likelihood ratios is described in [70]. In Program P4.3, the δ, δ', α and α' are denoted by del, delp, alp, and alpp. The first two distance measures dm(1) and dm(2) denote the ratios δ/α and δ'/α'.

We denote,

$$\Omega = \frac{1}{2}(\sigma/\sigma')^2(\delta/\alpha) + \frac{1}{2}(\sigma'/\sigma)^2(\delta'/\alpha') - 1 , \tag{4.5}$$

$$\cosh(\omega) - 1 = \Omega , \tag{4.6}$$

$$\omega = \ln[1 + \Omega + \sqrt{\Omega(2+\Omega)}] , \tag{4.7}$$

where ω is the cosh measure. The array element dm(3) is the distance measure 4.34294418ω, where ω is calculated for $\sigma = \sigma' = 1$ and the factor 4.34294418 is used to convert the cosh measure to decibels. The program `fin.m` converts Ω to the cosh measure dm(6). The parameters σ and σ' used above are the gain constants related to the cepstral gains by,

$$c_0 = \ln[\sigma^2] , \tag{4.8}$$

$$c_0' = \ln[(\sigma')^2] , \tag{4.9}$$

where the cepstral coefficients c_k and c_k' are the Fourier series coefficients for the log spectra given by,

$$\ln[\sigma^2/|A(e^{j\theta})|^2] = \sum_{k=-\infty}^{k=\infty} c_k e^{-jk\theta} , \tag{4.10}$$

$$\ln[(\sigma')^2/|A'(e^{j\theta})|^2] = \sum_{k=-\infty}^{k=\infty} c_k' e^{-jk\theta} . \tag{4.11}$$

```
% P4.3  - dist.m
function [ dm ] = dist( m, l, r, rp)
%   m          -         Filter order
%   l          -         Number of terms used in cepstral distance
%                        Measure
%   r          -         Autocorrelation sequence 1 (undistorted)
%   rp         -         Autocorrelation sequence 2 (distorted)
%   dm         -         Distances array

% DEFINE LOCAL CONSTANTS
DBFAC = 4.342944819;

% INITIALIZE LOCAL VARIABLES AND RETURN VECTORS
dm = zeros( 9, 1 );

% COMPUTE CEPSTRAL AND FILTER COEFFICIENTS
[ c, ra, alp, a, rc ] = cfind( m, l, r );
[ cp, rap, alpp, ap, rcp ] = cfind( m, l, rp );

% COMPUTE DM(0), DM(1)
del = ( r(1) * rap(1) ) + sum( 2 * r(2:m+1) .* rap(2:m+1) );
delp = ( rp(1) * ra(1) ) + sum( 2 * rp(2:m+1) .* ra(2:m+1) );
dm(1) = del / alp;
dm(2) = delp / alpp;

% COMPUTE DM(3)
q = ( ( dm(1) + dm(2) ) / 2.0 ) - 1;
if q >= 0.00
    dm(3) = fin(q);
end

% COMPUTE DM(4)
q1 = ( alpp * r(1) ) / ( alp * rp(1) );
q = ( ( ( dm(1) / q1 ) + dm(2) * q1 ) * 0.5 ) - 1.0;
if q >= 0.00
    dm(4) = fin(q);
end
```

Program P4.3: Computation of distance measures. (*Continues.*)

```
% COMPUTE DM(5)
q2 = alpp / alp;
q = ( ( ( dm(1) / q2 ) + dm(2) * q2 ) * 0.5 ) - 1.0;
if q >= 0.00
    dm(5) = fin(q);
end


% COMPUTE DM(6)
q = sqrt( dm(1) * dm(2) ) - 1.0;
qflag = ( q >= 0.00 );
if qflag
    dm(6) = fin(q);
end


% COMPUTE DM(7), DM(8), DM(9)
cepsum = 2 * sum( ( c(1:l) - cp(1:l) ) .^ 2 );
if cepsum >= 0.0
    dm(7) = DBFAC * sqrt( cepsum );
    q = log(q1);
    dm(8) = DBFAC * sqrt( cepsum + (q * q) );
    q = log(q2);
    dm(9) = DBFAC * sqrt( cepsum + (q * q) );;
else
    qflag = FALSE;
end
```

Program P4.3: (*Continued.*) Computation of distance measures.

To obtain the fourth distance measure dm(4), we consider the following values for σ^2 and $(\sigma')^2$,

$$\sigma^2 = \alpha/r(0), \tag{4.12}$$
$$(\sigma')^2 = \alpha'/r'(0), \tag{4.13}$$

where, r' indicates the autocorrelation lag corresponding to the quantized LSP frequencies. For dm(5), we take $\sigma^2 = \alpha$ and $(\sigma')^2 = \alpha'$. For the sixth distance measure, dm(6), the gain constants are adjusted to minimize Ω and the minimum is obtained as,

$$\Omega_{\min} = [(\delta/\alpha)(\delta'/\alpha')]^{1/2} - 1 . \tag{4.14}$$

The cepstral distance measure $u(L)$ is given by,

$$[u(L)]^2 = (c_0 - c_0')^2 + 2\sum_{k=1}^{L}(c_k - c_k')^2 , \tag{4.15}$$

where, L is the number of cepstral coefficients. The distance measure dm(7), is the cepstral distance calculated with $\sigma = \sigma' = 1$, dm(8) is the cepstral measure obtained with (4.12) and (4.13) holding true, and finally, dm(9) is the cepstral measure with the conditions $\sigma^2 = \alpha$ and $(\sigma')^2 = \alpha'$. The cepstral distance is a computationally efficient way to obtain the root-mean-square (RMS) difference between the log LPC spectra corresponding to the unquantized and quantized LSFs.

Example 4.3   The autocorrelation lags from Example P4.2 are used in Program P4.3 to find the distance measures. The order of the filter (m) is taken as 10 and the number of terms used to find the cepstral distance measure (l) is taken as 40. The distance measure array output from the program is given below.

$$dm = [1.0477;\ 1.0452;\ 1.3189;\ 1.7117;\ 1.7117;\ 1.3189;\ 1.2890;\ 1.6750;\ 1.6750]\,.$$

The cepstral distance dm(9) and likelihood ratios dm(1) and dm(2) for all frames from the speech file 'cleanspeech.wav' [76] are given in Figure 4.3. The frame energy and the spectrogram of the speech is also given in the same figure. The three distance measures follow the same trend over most of the frames, which is reasonable because of the fact that all three of them measure the distortion of spectra corresponding to quantized and unquantized LSFs. Note that no logical relationship can be drawn between the distance measures and the frame energy or the spectrogram of the speech signal.

4.4   SUMMARY

Procedures for computing three distinct types of spectral distance measures were presented in this chapter. These distance measures have a meaningful frequency domain interpretation [70]. The significance of the likelihood distance measures when the data are assumed to be Gaussian and their efficient computation using autocorrelation sequences are provided in a paper by Itakura [71]. The properties of cepstral coefficients from minimum phase polynomials [72] are exploited to compute the cepstral distance measure, which provides a lower bound for the rms log spectral distance. The cosh distance measure closely approximates the rms log spectral distance as well and provides an upper bound for it [70].

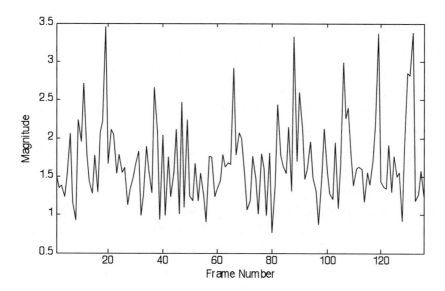

Figure 4.3: (a) Cepstral distance for all the frames.

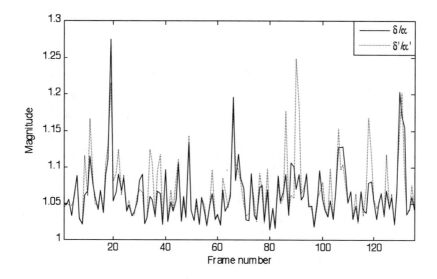

Figure 4.3: (b) Likelihood ratio distance measure at all the frames.

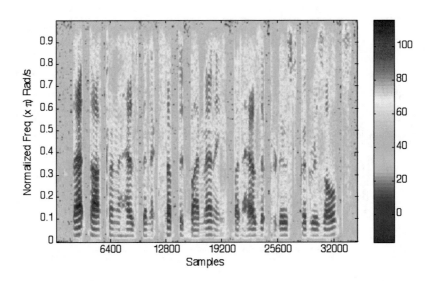

Figure 4.3: (c) Spectrogram at all the frames.

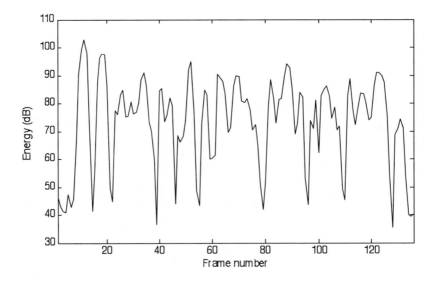

Figure 4.3: (d) Frame energy at all the frames.

CHAPTER 5

The Codebook Search

The prediction residual from the LP filter of the CELP analysis stage is modeled by an LTP and the Stochastic Codebook (SCB). The pitch delay of the LP residual is first predicted by the LTP and the SCB represents the random component in the residual. The LTP is also called as the Adaptive Codebook (ACB) because the memory of the LTP is considered as a codebook and the best matching pitch delay for a particular target is computed. Figure 5.1 provides an overview of the codebook search procedure in the analysis stage consisting of both the ACB and the SCB.

The codebook search procedure outlined in Figure 5.1 will be described in detail with the necessary theory and MATLAB programs in the following sections. Demonstration of some important details will also be provided with some numerical examples and plots.

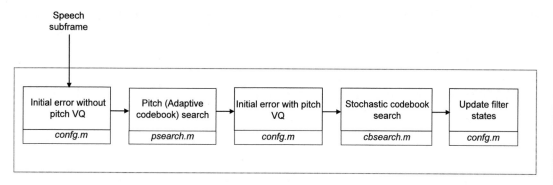

Figure 5.1: Codebook search procedure.

5.1 OVERVIEW OF CODEBOOK SEARCH PROCEDURE

The use of the closed-loop LTP was first proposed in a paper by Singhal and Atal [2] and led to a big improvement in the quality of speech. The vector excitation scheme using stochastic codevectors for A-by-S LP coding was introduced in a paper by Schroeder and Atal [1]. In this section, we give an overview of the structure of ACB and SCB and the search procedures. The codebook searches are done once per sub-frame for both the ACB and the SCB, using the LP parameters obtained from the interpolated LSPs of each sub-frame. Determining the best matching codevector for a particular signal is referred to as Vector Quantization (VQ) [73, 74]. The idea behind both the codebook searches is to determine the best match in terms of the codevector and the optimal gain

(gain-shape VQ [75]) so that the codevector is closest to the target signal in the MSE sense. Because of this reason, the theory and most of the MATLAB implementation of both the ACB and the SCB searches are similar. The major difference between the search procedures is that the target signal used in the process of error minimization is different. The target for the ACB search is the LP residual weighted by the perceptual weighting filter, $A(z/\gamma)$, where γ is the bandwidth expansion factor. The target for the SCB search is the target for the ACB search minus the filtered ACB VQ excitation [20]. Therefore, it is clear that the combination of the ACB and the SCB is actually attempting to minimize the perceptually weighted error. The entire search procedure is illustrated in Figure 5.2. i_a and i_s are the indices to the ACB and SCB that correspond to the optimal codevectors; g_a and g_s are the gains that correspond to the adaptive and stochastic codewords. In the MATLAB implementation, the ACB search is performed first and then the SCB search is performed.

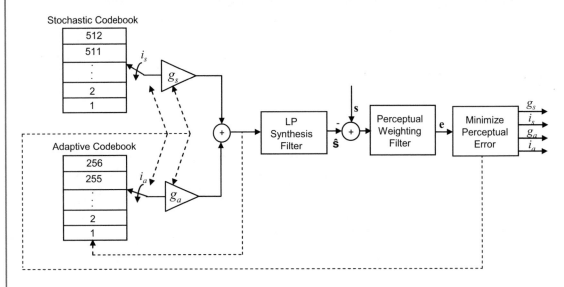

Figure 5.2: Searching the adaptive and the stochastic codebooks.

For the ACB, the codebook consists of codewords corresponding to 128 integer and 128 non-integer delays ranging from 20 to 147 samples. The search involves a closed-loop analysis procedure. Odd sub-frames are encoded with an exhaustive search requiring 8 bits whereas even sub-frames are encoded by a delta search procedure requiring 6 bits. Therefore, the average bits needed for encoding the ACB index is 7. Codebook search here means finding the optimal pitch delay and an associated gain that is coded between -1.0 and $+2.0$ using a modified Minimum Squared Prediction Error (MSPE) criterion. Sub-multiple delays are also checked using a sub-multiple delay table to check if the sub-multiples match within 1 dB of MSPE. This favors a smooth pitch contour and prevents pitch errors due to choosing of multiples of pitch delay. The adaptive codebook gets updated whenever the pitch memory of the LTP is updated, i.e., once every sub-frame.

For the fixed SCB, the codebook is overlapped with a shift of -2 samples between consecutive codewords. The codebook is 1082 bits long and consists of 512 codewords that are sparse and ternary valued $(-1, 0$ and $+1)$ [18, 46]. This allows for fast convolution and end correction of subsequent codewords, which will be described later in detail. A sample of two consecutive codewords shifted by -2 found in the stochastic codebook is given in Figure 5.3 where the overlap, sparsity and ternary values could be seen.

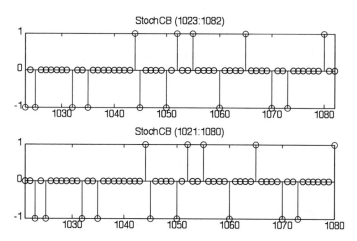

Figure 5.3: Examples of ternary valued SCB vectors.

As we mentioned earlier, the search procedures for both the ACB and the SCB are essentially the same, except that the targets are different. Assuming that we have the $L \times 1$ target vector $\mathbf{e}^{(0)}$ and codevectors $\mathbf{x}^{(i)}$, where i is the index of the codevector in the codebook, the goal is to find the index for which the gain score $g^{(i)}$ and the match score $m^{(i)}$ are maximized. The filtered codeword is given by,

$$\mathbf{y}^{(i)} = \mathbf{WHx}^{(i)} , \qquad (5.1)$$

where $\mathbf{W}$ and $\mathbf{H}$ are the $L \times L$ lower triangular matrices that represent the truncated impulse responses of the weighting filter and the LP synthesis filter, respectively. The overall product in (5.1) represents the effect of passing the codeword through the LP and the weighting filters. Figure 5.4 (adapted from [20]) illustrates the procedure to find the gain and the match scores. The scores are given by,

$$g^{(i)} = \frac{\mathbf{y}^{(i)\mathrm{T}}\mathbf{e}^{(0)}}{\mathbf{y}^{(i)\mathrm{T}}\mathbf{y}^{(i)}} , \qquad (5.2)$$

$$m^{(i)} = \frac{(\mathbf{y}^{(i)\mathrm{T}}\mathbf{e}^{(0)})^2}{\mathbf{y}^{(i)\mathrm{T}}\mathbf{y}^{(i)}} . \qquad (5.3)$$

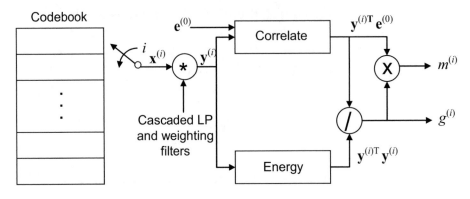

Figure 5.4: Finding gain and match scores [From [20]].

5.2 ADAPTIVE CODEBOOK SEARCH

Having looked at the match scores and the overview of codebook searches, we will now describe the ACB search in detail with MATLAB code. We will provide an elaborate study of the target signal for the ACB search, the integer delay search, and the sub-multiple/fractional delay search.

5.2.1 TARGET SIGNAL FOR ADAPTIVE CODEBOOK SEARCH

Denoting $\mathbf{s}$ and $\hat{\mathbf{s}}^{(0)}$ as the actual speech sub-frame vector and the zero input response of the LP filter, from Figure 5.2, we can see that the target signal for the ACB search, $\mathbf{e}^{(0)}$, should be,

$$\mathbf{e}^{(0)} = \mathbf{W}(\mathbf{s} - \hat{\mathbf{s}}^{(0)}) \ . \tag{5.4}$$

The filtered codeword $\mathbf{y}^{(i)}$ given in (5.1) is compared with the target $\mathbf{e}^{(0)}$ to get the gain and match scores as given in (5.2) and (5.3). The best codebook index i is the one that maximizes m_i.

The MATLAB implementation of the codebook search described above takes into account certain considerations in order to reduce the implementation complexity. From (5.1), we see that the codeword has to be filtered through the LP filter and the weighting filter for comparison with the target. One possible simplification is to do away with the LP filtering, so that we deal with the excitation signal instead of speech signal. Therefore, instead of using the weighting filter $A(z/\gamma)$, we use the inverse weighting filter $1/A(z/\gamma)$. The modified filtered codeword for comparison becomes,

$$\tilde{\mathbf{y}}^{(i)} = \mathbf{W}^{-1}\mathbf{x}^{(i)} \ . \tag{5.5}$$

This requires us to modify the target (converting it to a form of excitation) and the gain and match scores as well. These are given by,

$$\tilde{\mathbf{e}}^{(0)} = \mathbf{W}^{-1}\mathbf{H}^{-1}(\mathbf{s} - \hat{\mathbf{s}}^{(0)}) \, , \tag{5.6}$$

$$\tilde{g}_i = \frac{\tilde{\mathbf{y}}^{(i)\mathrm{T}}\tilde{\mathbf{e}}^{(0)}}{\tilde{\mathbf{y}}^{(i)\mathrm{T}}\tilde{\mathbf{y}}^{(i)}} \, , \tag{5.7}$$

$$\tilde{m}_i = \frac{(\tilde{\mathbf{y}}^{(i)\mathrm{T}}\tilde{\mathbf{e}}^{(0)})^2}{\tilde{\mathbf{y}}^{(i)\mathrm{T}}\tilde{\mathbf{y}}^{(i)}} \, . \tag{5.8}$$

```
% P5.1 - acbtarget.m
% s - Speech or residual segment
% l - Segment size
% d2 - Memory, 1/A(z)
% d3 - Memory, A(z)
% d4 - Memory, 1/A(z/gamma)
% no - LPC predictor order
% e0 - Codebook search initial error(zero vector) and updated
%      error(target)
% fc - LPC filter/weighting filter coefficients
% gamma2 - Weight factor, perceptual weighting filter

% LOAD THE SPEECH FRAME, FILTER MEMORIES AND OTHER PARAMETERS
load voic_fr

% INITIALIZE LOCALS
fctemp = zeros( no+1, 1 );
% COMPUTE INITIAL STATE, 1/A(z)
[ d2, e0 ] = polefilt( fc, no, d2, e0, l );
% RECOMPUTE ERROR
e0 = s - e0;
% COMPUTE INITIAL STATE, A(z)
[ d3, e0 ] = zerofilt( fc, no, d3, e0, l );
% COMPUTE INITIAL STATE, 1/A(z/gamma)
fctemp = bwexp( gamma2, fc, no );
[ d4, e0 ] = polefilt( fctemp, no, d4, e0, l );
```

Program P5.1: Target for the ACB.

The error is initialized to zero, and then it is passed as input to the LP synthesis filter $1/A(z)$. This gives the Zero Input Response (ZIR) i.e., the initial condition, of the LP synthesis filter. Then the ZIR is subtracted from the actual speech sub-frame and passed through the LP analysis filter, which converts it back to the residual. Since we work with prediction residuals now, the inverse weighting filter $1/A(z/\gamma)$ is used to obtain the final target vector for the ACB, i.e., $\tilde{\mathbf{e}}^{(0)}$. Program P5.1 performs this operation, but it needs supporting files and workspace [76].

Example 5.1   A test speech segment (voiced) is taken and tested with Program P5.1. The segment of speech and the target are shown in Figure 5.5. Note that this target vector will be used by the ACB to predict the pitch period.

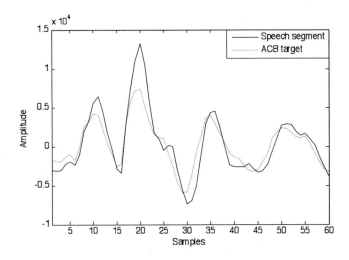

Figure 5.5: Speech segment and target for the ACB.

5.2.2   INTEGER DELAY SEARCH

Once the target signal is generated, the next step is to search for the integer pitch delay. In the MATLAB implementation, the pitch is coded in two stages – searching the integer delays first and the neighboring non-integer delays next. In this section, we will look into the coding of integer pitch delays.

The MATLAB program that implements the integer pitch delay (`integerdelay.m`) is available under the directory P5_2 in the website [76]. Initially, the excitation vector (`v0`) is loaded with a copy of the pitch memory (`d1b`), and allowable search ranges are fixed in the variables `minptr` and `maxptr`. If the sub-frame is odd, then a full search of the pitch delay table (`pdelay`) is done. If the sub-frame is even, the pitch delay is searched between −31 to 32 of the pitch index in the pitch delay table. The pitch delay table has both integer and fractional pitch values. In the integer delay search, only the integer delays are searched and match score is set to zero, if the pitch delay is fractional.

The gain and match scores are computed as given in (5.7) and (5.8) in the MATLAB code `pgain.m` and portions of it are provided in Program P5.3. The full MATLAB code for computing the

```
% RUN GAIN COMPUTATIONS
if first == 1
    % CALCULATE AND SAVE CONVOLUTION OF TRUNCATED (TO LEN)
    % IMPULSE RESPONSE FOR FIRST LAG OF T (=MMIN) SAMPLES:
    %
    %             MIN(i, len-1)
    %       y    =  SUM   h * ex       , WHERE i = 0, ..., L-1 POINTS
    %      i, t     j=0    j   i-j
    %
    %                         h |0 1...len-1 x x|
    %       ex |L-1  . . .  1 0|           = y[0]
    %          ex |L-1  . . .  1 0|        = y[1]
    %                          :              :
    %                       ex |L-1  . . .  1 0| = y[L-1]
    for i = 0:l-1
        jmax = min( i, len-1 );
        Ypg(i+1) = sum( h(1:jmax+1) .* ex(i+1:-1:i-jmax+1) );
    end
else
    % END CORRECT THE CONVOLUTION SUM ON SUBSEQUENT PITCH LAGS
    %       y    = 0
    %      0, t
    %      y    =  y         + ex  * h   WHERE i = 1, ..., L POINTS
    %      i, m     i-1, m-1    -m   i   AND    m = t+1, ..., tmax LAGS
    Ypg(len-1:-1:1) = Ypg(len-1:-1:1) + ( ex(1) * h(len:-1:2) );
    Ypg(l:-1:2) = Ypg(l-1:-1:1);
    Ypg(1) = ex(1) * h(1);
end
% FOR LAGS (M) SHORTER THAN FRAME SIZE (L), REPLICATE THE SHORT
% ADAPTIVE CODEWORD TO THE FULL CODEWORD LENGTH BY OVERLAPPING
% AND ADDING THE CONVOLUTION
y2(1:l) = Ypg(1:l);
if m < l
    % ADD IN 2ND CONVOLUTION
    y2(m+1:l) = Ypg(m+1:l) + Ypg(1:l-m);
    if m < fix(l/2)
        % ADD IN 3RD CONVOLUTION
        imin = ( 2 * m ) + 1;
        imax = l;
        y2( imin:imax ) = y2( imin:imax ) + Ypg( 1:l-(2*m) );
    end
end
```

Program P5.3: Gain and match scores computation. (*Continues.*)

```
% CALCULATE CORRELATION AND ENERGY
cor = sum( y2(1:l) .* e0(1:l) );
eng = sum( y2(1:l) .* y2(1:l) );

% COMPUTE GAIN AND MATCH SCORES
if eng <= 0.0
    eng = 1.0;
end
Pgain = cor / eng;
match = cor * Pgain;
```

Program P5.3: (*Continued.*) Gain and match scores computation.

gain and match scores are available in the website under the directory P5_3 [76]. In this program, Ypg is the variable that stores the filtered codeword. The filtered codeword is obtained by convolving the excitation (ex) with the truncated impulse response (h) of the inverse weighting filter. The convolution is shown in the first if condition in Program P5.3. Full convolution is performed only when the first delay is tested. Because of the integer delay, the excitation signal for the second delay is simply shifted by one from the excitation signal for the first delay. Exploiting this, we can perform just end corrections from the second convolution onwards and reduce the number of computations. This is illustrated in the first else condition in Program P5.3. The newly arrived first sample in the excitation is multiplied with the time-reversed impulse response and added to the previous convolution result. The new result is restricted to the length of the excitation signal (60).

If the pitch delay (M) is greater than the sub-frame length (60), then the codeword is constructed by taking the previous excitation signal and delaying it by M samples. If M is less than 60, then the short vector obtained by delaying the previous excitation signal is replicated by periodic extension. In Program P5.3, the effect of periodic extension is provided by overlap-add of the convolution result. Once the filtered codeword is obtained, the gain and match scores are computed exactly in the same way as in (5.7) and (5.8). For even sub-frames the same procedure is repeated except that the search range is limited.

Example 5.2   Integer delay search is performed for the speech segment given in Example 5.1. The sub-frame is odd; therefore, all the integer delays are searched. The gain and match scores with respect to the indices of the pitch delay table are plotted as shown in Figure 5.6. The maximum index is 117 and the actual pitch period is 56. This is obtained by running the MATLAB program integerdelay.m.

5.2.3   SUB-MULTIPLE/FRACTIONAL DELAY SEARCH

After performing an integer delay search, 2, 3, and 4 sub-multiple delays are searched according to the sub-multiple delay table (submult), in order to ensure that the pitch contour is smooth. Note

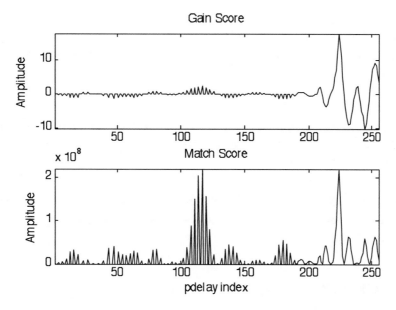

Figure 5.6: Gain and match scores for ACB search in a voiced segment.

that only the sub-multiples specified in the sub-multiple delay table are searched. Sub-multiple pitch delay is selected if the match is within 1dB of MSPE of the actual pitch delay already chosen.

We will now describe fractional pitch estimation, which involves the interpolation of the codevectors (that correspond to integer delay) chosen from the pitch memory (ACB). After the interpolation is performed and codevectors corresponding to the fractional delay are found, the gain and match scores are determined as shown before (pgain.m). The interpolating function is the product of a hamming window (h_w), and a sinc function and totally 8 points are used for interpolation. If D fractional samples are to be computed between two integer samples, it is equivalent to increasing the sampling frequency by D. Therefore, the signal is upsampled and low-pass filtered, which increases the number of samples by a factor of D. Then, the oversampled signal is delayed by an integral amount equivalent to the fractional delay of the original signal. Again, the signal is downsampled to the actual sampling rate. The entire process can be replicated using only windowed sinc interpolation in the time domain [17]. The window and the windowed interpolation functions are given as,

$$h_w(k) = 0.54 + 0.46 \cos\left(\frac{k\pi}{6N}\right) , \tag{5.9}$$

where N = interpolation length (even) $k = -6N, -6N + 1, \ldots, 6N$ and

$$w_f(j) = h(12(j + f))\frac{\sin((j + f)\pi)}{(j + f)\pi} , \tag{5.10}$$

where $j = \frac{-N}{2}, \frac{-N}{2} + 1, \ldots, \frac{N}{2} - 1$, $f = \frac{1}{4}, \frac{1}{3}, \frac{1}{2}, \frac{2}{3}, \frac{3}{4}$ (fractional part).

Program P5.4 (fracdelay.m) implements fractional delaying of an input signal and it is available under the directory P5_4 in the website [76]. After computing the windowed sinc function, the input signal is multiplied with the windowed sinc and interpolated to find the signal delayed by fractional amount. The delayed excitation signal is used in the next step to find the gain and the match score.

The integer and fractional parts of the delay are computed and the optimal excitation vector with the highest match score is found. This excitation vector is used to update the pitch memory, so that the codebook search for the next sub-frame uses the updated codebook.

Example 5.3 The windowed interpolating functions (8 points) for the fractional delays for 1/4 and 3/4 are shown in Figure 5.7. A signal and its delayed version by a delay of 1/3 are shown in Figure 5.8. These figures are generated from the program fracdelay.m.

5.3 STOCHASTIC CODEBOOK SEARCH

The SCB search is similar to the ACB search except that the SCB is stored as overlapped codewords, and the target signal for the SCB search is different to that of the ACB search. The following sections describe the generation of target signals for SCB search and the actual search procedure.

5.3.1 TARGET SIGNAL FOR STOCHASTIC CODEBOOK SEARCH

The SCB search is done on the residual signal after the ACB search identifies the periodic component of the speech signal. If $\mathbf{u}$ is the codeword obtained from the ACB search, the target for the SCB search is,

$$\mathbf{e}^{(0)} = \mathbf{W}(\mathbf{s} - \hat{\mathbf{s}}^{(0)}) - \mathbf{WHu} . \tag{5.11}$$

The error target for the SCB is the error target for the ACB subtracted from the filtered codeword obtained after the ACB search. In the MATLAB implementation, instead of operating directly with the error target, the target signal is converted into a form of excitation as done with the ACB search. Therefore, the modified filtered codeword for comparison is same as the one given in (5.5). In this case, $\mathbf{x}^{(i)}$, is the ith SCB codeword used for comparison. The modified target is given by,

$$\tilde{\mathbf{e}}^{(0)} = \mathbf{W}^{-1}\mathbf{H}^{-1}\mathbf{W}^{-1}\mathbf{e}^{(0)} , \tag{5.12}$$

which is the same as,

$$\tilde{\mathbf{e}}^{(0)} = \mathbf{W}^{-1}\mathbf{H}^{-1}(\mathbf{s} - \hat{\mathbf{s}}^{(0)}) - \mathbf{W}^{-1}\mathbf{u} . \tag{5.13}$$

The gain and match scores are also the same as given in Equations (5.7) and (5.8). The pitch synthesis routine that generates $\mathbf{u}$ from the pitch delay and gain parameters obtained from the ACB search is given in the Program P5.5 (pitchsynth.m) available under the directory P5_5 in the website [76].

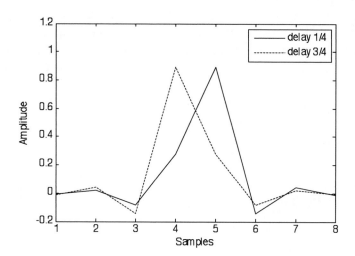

Figure 5.7: Windowed sinc function for interpolation.

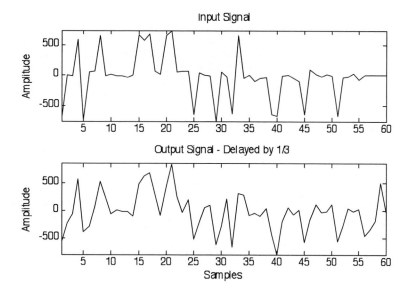

Figure 5.8: Input signal and its delayed version by 1/3 sample.

The variable buf is the memory that contains the past excitation. The array b contains the pitch delay and the pitch predictor coefficients. The integer and fractional parts of the delay are obtained from the pitch delay. If the integer part of pitch delay (M_i) is greater than or equal to the frame size (60 samples), the current excitation in the pitch memory is updated with the M_i samples of the past excitation stored in the pitch memory. If M_i is less than 60 samples, the pitch memory is updated by extending the past excitation to 60 samples. Once the integer delaying is performed, fractional delay is implemented by the program ldelay.m. The current ACB excitation is now present in buf, which is scaled by the pitch gain to obtain the scaled ACB excitation vector **u** and stored in rar. For $M > 60$, this procedure is equivalent to doing pitch synthesis using a long-term predictor with delay M and appropriate gain.

```
% P5.6 - scbtarget.m
%   s              -        Speech or residual segment
%   l              -        Segment size
%   d1             -        Memory, 1/P(z)
%   d2             -        Memory, 1/A(z)
%   d3             -        Memory, A(z)
%   d4             -        Memory, 1/A(z/gamma)
%   fctemp         -        Bandwidth expanded poles, 1/A(z/gamma)
%   idb            -        Dimension of d1a and d1b (209)
%   no             -        LPC predictor order (10)
%   bb             -        Pitch predictor coefficients (array of 4)
%   e0             -        Codebook search initial error and updated error
%   fc             -        LPC filter/weighting filter coefficients
%   gamma2         -        Weight factor, perceptual weighting filter

load scbvect;

% INITIALIZE LOCALS
fctemp = zeros( MAXNO+1, 1 );
% COMPUTE INITIAL STATE, 1/P(z)
[ e0, d1 ] = pitchvq( e0, l, d1, idb, bb, 'long' );]
% COMPUTE INITIAL STATE, 1/A(z)
[ d2, e0 ] = polefilt( fc, no, d2, e0, l );
```

Program P5.6: Generating the SCB target vector.

After generating **u**, the target vector $\tilde{\mathbf{e}}^{(0)}$ needs to be computed. This is performed in Program P5.6 (scbtarget.m). The function pitchvq.m is very similar to Program P5.5 and generates the vector **u**. Filtering **u** with $1/A(z)$ is the same as computing $\mathbf{Hu} + \hat{\mathbf{s}}^{(0)}$, where $\hat{\mathbf{s}}^{(0)}$ is the ZIR of $1/A(z)$. This is subtracted from **s** to obtain the residual $\mathbf{s} - \hat{\mathbf{s}}^{(0)} - \mathbf{Hu}$. This residual is passed through the LP analysis filter $A(z)$ and the inverse weighting filter $1/A(z)$ to yield the vector $\mathbf{W}^{-1}\mathbf{H}^{-1}(\mathbf{s} - \hat{\mathbf{s}}^{(0)}) - \mathbf{W}^{-1}\mathbf{u}$, which is the same as $\tilde{\mathbf{e}}^{(0)}$, the error target, in (5.13).

Example 5.4 The scaled ACB excitation is generated for the speech segment in Example 5.1 are computed using Program P5.5. The workspace variable `pitsyn` contains the necessary pitch delay and gain parameters needed by the program. The excitation is plotted in Figure 5.9.

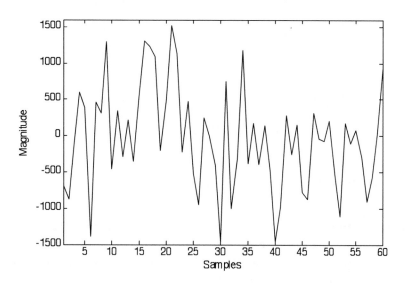

Figure 5.9: The scaled ACB excitation.

Example 5.5 The SCB error target is generated using Program P5.6 and is shown in Figure 5.10. Note that the target is given by $\tilde{\mathbf{e}}^{(0)}$ and not $\mathbf{e}^{(0)}$. Therefore, the target displayed has the form of excitation and needs to be filtered with $1/A(z)$, if we want to see the actual error target $\mathbf{e}^{(0)}$. This could be easily determined using the function `polefilt.m`.

5.3.2 SEARCH FOR BEST CODEVECTOR

The SCB search is done with the modified error target signal. This is very similar to the ACB search except that the SCB codebook is fixed and has some structure that could be exploited to perform fast convolutions. Details on the structure of the SCB are given in Section 5.1.

Program P5.7 (`cgain.m`) computes the gain and matches scores for a particular SCB excitation. It is available in the directory P5_7 in the website [76]. For the first search alone, a full convolution is performed to compute `Ycg`, the filtered codeword. Filtering is performed by convolving the excitation (`ex`) with the truncated impulse response (`h`) of the inverse weighting filter. The

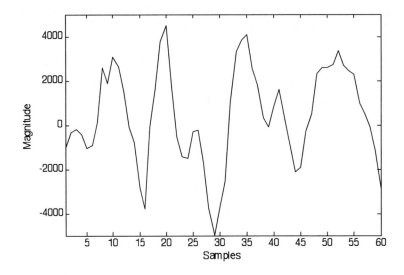

Figure 5.10: Example of a modified SCB target signal.

full convolution is performed in the first `if` structure in the program. Since the subsequent code-words are shifted by −2, the `else` structure corrects the previous convolution result individually for the two shifts. The gain and match scores are then computed using the filtered codeword. This process is repeated for all possible codewords, and the one with the best match score is chosen. The gain and match scores are recomputed with the chosen codeword to correct any errors accumulated during the process. The selected codeword is scaled with the quantized gain in order to get the SCB excitation.

After determining the SCB excitation, the filter states are updated using a procedure similar to the one followed in Program P5.6. The difference is that the error $\tilde{\mathbf{e}}^{(0)}$ is not initialized to a zero vector but the SCB excitation itself. This results in the complete update of the states of the filters given in Figure 5.2. The final update of filter states is performed on the synthesis part of the A-by-S procedure.

Example 5.6 The gain and the match scores are computed for the SCB target given in Example 5.5. The resulting scores are plotted against the indices of the SCB codewords in Figure 5.11. The gain scores are quantized gain values and they are constrained to the limits imposed by the modified excitation gain scaling procedure [20]. In this case, the best codebook index is 135 and the corresponding quantized gain score is −1330.

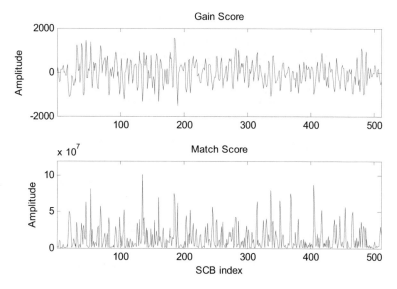

Figure 5.11: Gain and match scores for SCB search in a voiced segment.

5.4   BITSTREAM GENERATION

The bitstream generation process for the FS-1016 encoder consists of packing the quantized LSF values, the pitch delay, the pitch gain, the stochastic codeword and the stochastic gain indices into a bitstream array and adding bit error protection to the bitstream. The total 144 bits per frame are allocated as shown in Table 5.1.

The total number of bits needed to encode the quantized LSF values per frame is 34. Though LSFs of sub-frames are used for codebook searches, LSFs are transmitted on only a per-frame basis. This is because the sub-frame LSFs could be easily recovered from the LSFs of a frame using linear interpolation. The ACB pitch delay index is encoded with 8 bits for odd sub-frames and 6 bits for even sub-frames. Recall that the ACB pitch delay for odd sub-frames is absolute and the pitch delay for even sub-frames is relative with respect to the previous odd sub-frame. Only the absolute pitch delays have 3 bits protected by the FEC since coding the absolute pitch delays is crucial for coding the delta delays correctly. The ACB gains are encoded using 5 bits per sub-frame, and therefore, 20 bits are needed to encode the ACB gain values per frame. The ACB gain has its most perceptually sensitive bit protected. The SCB indices need 9 bits per sub-frame and the SCB gains need 5 bits per sub-frame. Therefore, 56 bits are used per frame for the SCB.

One bit per frame is used for synchronization, 4 bits per frame for Forward Error Correction (FEC) and 1 bit is used for future expansion. The FEC consists of encoding selected bits using a Hamming code. The Hamming code used is a (15,11) code which means that 11 bits are protected using 4 bits to give a total of 15 bits. Hamming codes can detect 2 errors and correct 1 error [77].

	Table 5.1: Bit allocation for a frame.				
	Sub-frame				
Parameter	**1**	**2**	**3**	**4**	**Frame**
ACB index	8	6	8	6	28
ACB gain	5	5	5	5	20
SCB index	9	9	9	9	36
SCB gain	5	5	5	5	20
LSF 1					3
LSF 2					4
LSF 3					4
LSF 4					4
LSF 5	Sub-frames not				4
LSF 6	applicable here				3
LSF 7					3
LSF 8					3
LSF 9					3
LSF 10					3
Future Expansion					1
Hamming Parity					4
Synchronization					1
Total					144

3 bits of the pitch delay index for the odd sub-frames and 1 bit of the pitch gain for all the sub-frames are protected using the Hamming code. The 11th bit protected by the Hamming code is the future expansion bit.

5.5   SUMMARY

In this chapter, we discussed the structure of the ACB and the SCB and search methods to find an optimal codevector. Further improvements in the codebook structure were made with the use of algebraic codes for the SCB, and the modified CELP algorithm came to be known as ACELP. The reduction in complexity was obtained using backward filtering and algebraic codes such as Reed-Muller code and Nordstrom-Robinson code [33]. Most of the modern CELP algorithms, such as the G.729 and the SMV are categorized as Conjugate Structured ACELP (CS-ACELP) [4] algorithms. They use Conjugate Structure-VQ (CS-VQ) to perform joint quantization of the adaptive and the stochastic excitation gains [49]. In AMR-WB, multiple bit rates are achieved using algebraic codebooks of different sizes and encoding the LP parameters at different rates. Furthermore, the pitch and algebraic codebook gains are jointly quantized in the AMR-WB [5].

CHAPTER 6

The FS-1016 Decoder

The parameters of the encoded bitstream are decoded in the FS-1016 CELP receiver. The decoded parameters include the ACB and the SCB gains and indices for each sub-frame. Also included in the decoded parameters are the LSFs of a frame, which are then interpolated to obtain sub-frame LSFs. The SCB gain, the ACB gain and the pitch delays are smoothed in case transmission errors are detected. Composite excitation with both the periodic and stochastic components is generated for the LP synthesis filter. The raw synthesized speech is scaled and clamped to the 16-bit integer range, and the distance measures between input and synthesized speech are computed. The synthesized speech is postfiltered to enhance the formant structure of speech, and it is also filtered using a high-pass filter. Therefore, there are three speech outputs from the CELP receiver: raw synthesized (non-postfiltered), postfiltered and high-pass filtered speech.

The block diagram of the CELP receiver is illustrated in Figure 6.1 and the key functions will be described in the following sections. Demonstration of the functions will be provided using MATLAB programs and plots.

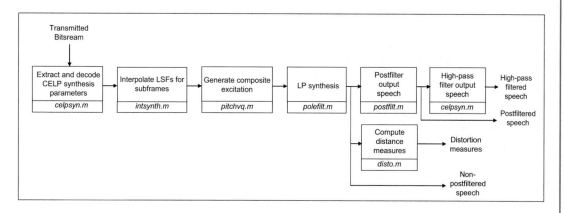

Figure 6.1: The CELP receiver.

6.1   EXTRACTING PARAMETERS FOR SYNTHESIS

The CELP synthesis parameters of a particular frame are extracted and decoded from the bitstream as the first step in the CELP synthesis process. This step also involves decoding the LSFs using the quantization table and interpolating them to create sub-frame LSPs. Smoothing of the ACB, SCB gains and the pitch delays is also performed as a part of this process.

The codewords are extracted from the received bitstream and the Hamming error protected bitstream is decoded. The bit error rate is computed as a running average of bad syndromes. In Hamming code, a syndrome other than zero indicates the position of error and hence is called a bad syndrome whereas zero indicates no error [77]. The LSFs of the particular frame are extracted, decoded and interpolated to create sub-frame LSFs. The LSFs are decoded by indexing the quantization table using the transmitted indices to create quantized LSFs.

The quantized values of LSFs are then used to create sub-frame LSFs using linear interpolation as shown in Program P6.1. The quantized LSFs are tested for monotonicity again in the synthesis stage. Corrections are done similar to the analysis stage to make the LSFs monotonic if they are non-monotonic. Sub-frame LSFs are computed by interpolating the LSFs of the current and the previous frames.

```
% P6.1 - synthLSP.m
%    nn          -      Number of subframes per frame
%    lspnew      -      New frequency array (LSPs)
%    no          -      Number of LSPs
%    lsp         -      Interpolated frequency matrix


load synLSP_data;
% CHECK LSPs; TRY TO FIX NONMONOTONIC LSPs BY REPEATING PAIR
if any( lspnew(2:no) <= lspnew(1:no-1) )
    for i = 2:no
        if lspnew(i) <= lspnew(i-1)
            lspnew(i) = lspoldIS(i);
            lspnew(i-1) = lspoldIS(i-1);
        end
    end
end
% RECHECK FIXED LSPs
if any( lspnew(2:no) <= lspnew(1:no-1) )
    % REPEAT ENTIRE LSP VECTOR IF NONMONOTONICITY HAS PERSISTED
    lspnew = lspoldIS;
end
% INTERPOLATE LSPs AND TEST FOR MONOTONICITY
for i = 1:nn
    lsp( i, 1:no ) = ( ( wIS(1,i) * lspoldIS( 1:no ) ) + ...
                       ( wIS(2,i) * lspnew( 1:no ) ) )';
    % CHECK FOR MONOTONICALLY INCREASING LSPs
    if any( lsp(i,2:no) <= lsp(i,1:no-1) )
        fprintf( 'intsynth: Nonmonotonic LSPs @ the current frame \n');
    end
end
```

Program P6.1: Interpolating LSFs for sub-frames.

The weights used for the interpolation are given in Table 6.1. Indicating the LSF of the previous frame as f_o and the LSF of the current frame as f_n, the LSFs of the ith sub-frame are computed as follows,

$$f_{i,s} = w_{i,o} f_o + w_{i,n} f_n \,, \tag{6.1}$$

$w_{i,o}$ is the weight corresponding to the ith sub-frame LSF that multiplies the previous frame LSF and $w_{i,n}$ is the weight corresponding to the ith sub-frame LSF that multiplies the current frame LSF.

Table 6.1: Interpolation weights for sub-frame LSFs.

	Sub-frame Number			
	1	2	3	4
Previous Frame	0.8750	0.6250	0.3750	0.1250
Current Frame	0.1250	0.3750	0.6250	0.8750

The ACB and the SCB parameters are decoded from the bitstream. Smoothing is performed on the SCB gain, the ACB gain and the pitch delay if bit protection is enabled. Smoothing is essentially the process of reducing the variations in the received parameters across sub-frames. Here, we will illustrate the case of gain smoothing and the smoothing of pitch delay is also done in a similar manner. Gain smoothing is performed only when errors are detected. This is because, when the actual gain is not correct, the smoothed gain from the past and future sub-frames will be a good estimate. If either two errors are detected or the bit error rate is more than permissible, gain smoothing is initiated. This is done by creating a vector of past and future gain values. The mean and variance of the gain values in the vector are estimated. If the variance is within the prescribed limit and the gain value of the current sub-frame falls outside the permissible range, the gain of the current sub-frame is set to the average gain value, retaining the actual sign. For the fourth sub-frame in any frame, gain smoothing is disabled.

Example 6.1   An illustration of generating sub-frame LSFs from the LSFs of a frame is described in this example. The sub-frame LSFs are obtained from the LSFs of the previous frame and the current frame using Program P6.1. Table 6.2 shows the values of LSFs for the current frame, previous frame and the four sub-frames in the sub-frame analysis buffer. The pole-zero plots of the LP synthesis filters corresponding to the LSFs of the previous frame, the LSFs of the current frame, and the interpolated LSFs of the sub-frames are given in Figure 6.2. The effect of interpolating the LSFs of the previous and the current frames is clearly visible in the figure. The pole-zero plot corresponding to the LSF of the first sub-frame is closest to that of the previous frame whereas the pole-zero plot corresponding to the LSF of the fourth sub-frame is closest to that of the current frame.

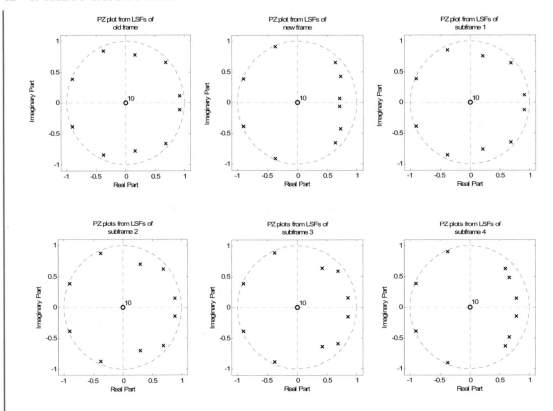

Figure 6.2: Pole-zero plots from LSFs (a) for the previous frame, (b) for the current frame, (c) for the sub-frame 1, (d) for the sub-frame 2, (e) for the sub-frame 3 and (f) for the sub-frame 4.

6.2   LP SYNTHESIS

The SCB excitation is generated by scaling the selected SCB codeword by the SCB gain value. The composite excitation is generated by passing the SCB excitation through the LTP and Program P5.5 is used to perform this. This excitation will be used in the LP synthesis filter to synthesize the speech signal. The LP synthesis filter is an all-pole filter and the denominator polynomial is computed by converting the sub-frame LSFs to the coefficients of the LP polynomial $A(z)$ using (3.3). Program P4.1 is used to convert the sub-frame LSFs to the corresponding LP coefficients. The LP coefficients and the composite excitation are used in the all-pole synthesis filter to synthesize the speech signal.

The synthesized speech is scaled and clamped to the 16-bit range. Program P6.2 generates the raw synthesized (non-postfiltered) speech from the LSFs of a sub-frame and the excitation. It internally uses the Programs P5.5 and P4.1 in order to generate the composite excitation and convert the LSFs to direct-form LP coefficients.

Table 6.2: Interpolating LSFs of a frame to create sub-frame LSFs.

Frame LSF		Sub-frame LSF			
Old	New	1	2	3	4
0.0213	0.0350	0.0230	0.0264	0.0298	0.0333
0.0600	0.0700	0.0612	0.0638	0.0663	0.0688
0.1187	0.0969	0.1160	0.1105	0.1051	0.0996
0.1350	0.1213	0.1333	0.1298	0.1264	0.1230
0.1988	0.1412	0.1916	0.1772	0.1628	0.1484
0.2500	0.2288	0.2473	0.2420	0.2367	0.2314
0.3100	0.3100	0.3100	0.3100	0.3100	0.3100
0.3312	0.3156	0.3293	0.3254	0.3215	0.3176
0.4288	0.4288	0.4288	0.4288	0.4288	0.4288
0.4363	0.4363	0.4363	0.4363	0.4363	0.4363

Example 6.2   The composite excitation is generated and LSFs are converted to LP coefficients for sub-frame 3 of Example 6.1. The excitation given in Figure 6.3 and LP coefficients are used to synthesize speech given in Figure 6.4.

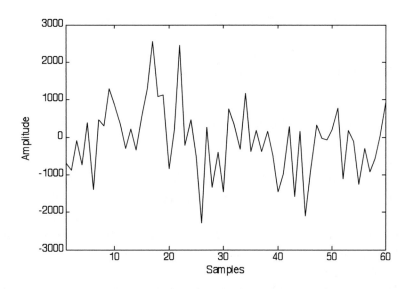

Figure 6.3: Composite excitation with both periodic and stochastic components.

```
% P6.2 - synth_speech.m
% subframe    - Subframe number in the given frame
% scaling     - Scaling factor for output speech
% stoch_ex    - Stochastic excitation
% comp_ex     - Composite excitation with adaptive and stochastic
%                 components
% lsp         - Matrix with LSPs of 4 subframes, each being a single row
% pc          - LP coefficients for a particular subframe
% npf         -  Non-postfiltered speech subframe

load excit_param;
subframe=3;
scaling=1;
maxv=ones(1,60)*32767;
minv=ones(1,60)*-32768;

% GENERATE PITCH EXCITATION VECTOR AND COMBINE WITH STOCHASTIC
% EXCITATION TO PRODUCE COMPOSITE LPC EXCITATION VECTOR
[comp_ex dps] = pitchsynth( stoch_ex, 1, dps, idb, bb, 'long' );
% CONVERT THE LSP OF A SUBFRAME TO CORRESPONDING PC
pc = lsptopc( lsp(subframe,:), no );
% SYNTHESIS OF NON-POSTFILTERED SPEECH
[ dss,npf  ] = polefilt( pc, no, dss, comp_ex, 1 );

% SCALE AND CLAMP SYNTHESIZED SPEECH TO 16-BIT INTEGER RANGE
npf = npf * descale;
npf = min( [ npf'; maxv ] )';
npf = round( max( [ npf'; minv ] )' );
```

Program P6.2: Synthesizing speech from excitation and LSF.

6.3   POSTFILTERING OUTPUT SPEECH

Postfiltering of output speech is generally performed in order to improve the perceptual quality of the output speech [47]. Noise levels are kept as low as possible in the formant regions during encoding as it is quite difficult to force noise below the masking threshold at all frequencies. The major purpose for postfiltering is to attenuate the noise components in the spectral valleys, which was not attenuated during encoding.

Postfilters can be generally designed by moving the poles of the all-pole LP synthesis filter thereby resulting in filters of the form $1/A(z/\alpha)$. But, using this filter to achieve sufficient noise reduction will result in severe muffling of speech [47]. This is because the LP synthesis filter has a low-pass spectral tilt for voiced speech. Therefore, the spectral tilt of the postfilter must be compensated to

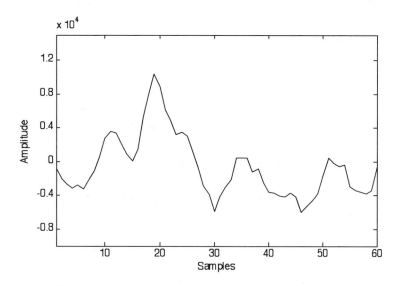

Figure 6.4: Synthesized (non-postfiltered) speech using the composite excitation and LP coefficients.

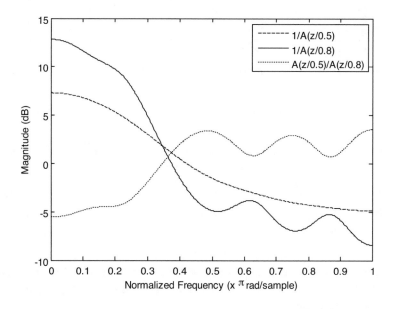

Figure 6.5: Frequency response of the pole-zero postfilter.

avoid muffling. Figure 6.5 illustrates the frequency response magnitude of $1/A(z/\alpha)$ for the case of $\alpha = 0.5$ and $\alpha = 0.8$. $A(z)$ is computed for a voiced frame, and it is instructive to note that $\alpha = 0.5$ gives the spectral tilt alone and $\alpha = 0.8$ gives the filter with both the spectral tilt and the formant structure. Therefore, the spectral tilt can be removed using the filter of form $A(z/0.5)/A(z/0.8)$, which is the pole-zero postfilter where the spectral tilt is partially removed and is also shown in Figure 6.5.

Further compensation for the low-frequency tilt and muffling can be provided by estimating the spectral tilt (first-order fit) of the postfilter denominator polynomial. This is done by computing the reflection coefficients of the denominator polynomial. The first reflection coefficient is the LP coefficient $a_1(1)$ of the first-order polynomial that fits the spectrum. An all-zero filter of the form $1 + \mu a_1(1)z^{-1}$ with $\mu = 0.5$ is used for the adaptive compensation of spectral tilt. The response of this filter and the overall response of the pole-zero postfilter with adaptive spectral tilt compensation are given in Figure 6.6.

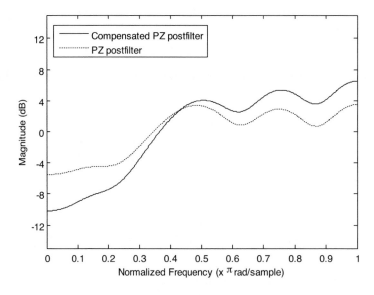

Figure 6.6: Pole-zero postfilter with and without adaptive spectral tilt compensation.

The power of the synthesized speech before postfiltering (input power) and the after postfiltering (output power) are computed by filtering the squared signal using an exponential filter. Exponential filters are generally used in time series analysis for smoothing the signals and are described by the transfer function,

$$H_s(z) = \frac{\tau}{1 - (1-\tau)z^{-1}}, \qquad (6.2)$$

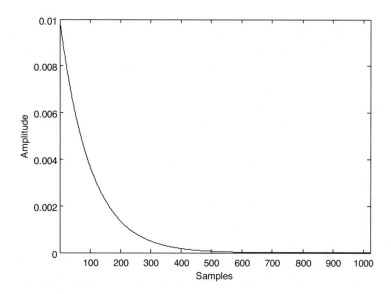

Figure 6.7: Impulse response of the exponential smoothing filter.

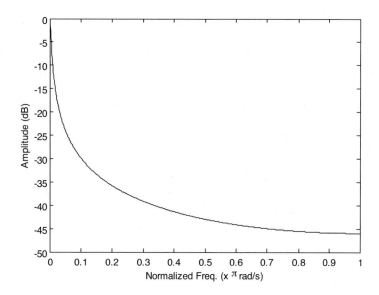

Figure 6.8: Magnitude frequency response of the exponential smoothing filter.

where τ is the parameter that controls the degree of smoothing; smaller the parameter τ, greater the degree of smoothing. The smoothed average estimates of the input and the output power of a sub-frame are used for Automatic Gain Control (AGC). The impulse and frequency responses of the exponential filter with $\tau = 0.01$ are shown in Figures 6.7 and 6.8, respectively. Note that since the filter is IIR, only the truncated impulse response is shown. Because of the properties of the Fourier transform, the magnitude response of the exponential filter is also a decaying exponential function.

The purpose of the AGC is to scale the output speech such that its power is roughly the same as unfiltered noisy output speech. The scaling factor is computed as,

$$S = \sqrt{\frac{P_{unf}}{P_{out}}}, \tag{6.3}$$

where P_{unf} and P_{out} are the unfiltered speech power and output speech power, respectively. The output speech is then scaled using the scaling factor. The MATLAB code for adaptive postfiltering and AGC is given in Program P6.3.

The two different values of α used in the pole-zero postfilter are stored in the variables `alpha` and `beta` in Program P6.3. The filter memories of the pole and zero filters are also taken into consideration during postfiltering. All-zero filtering is also performed to provide further compensation to spectral tilt. The negative of the first reflection coefficient is taken as $a_1(1)$ because of the sign convention used in the program. The input and output power estimates are computed using exponential filters taking into account the memory of the filters as well.

Example 6.3   Postfiltering is illustrated for the sub-frame 3 of the speech segment given in Example 6.1. Figure 6.9 shows the plot of the postfiltered and scaled output speech sub-frame. Figure 6.10 shows the plot of scaling factors from the AGC module for the speech sub-frame. For this particular sub-frame, the scaling factors are less than 1 and the postfiltered speech has to be scaled down.

6.4   COMPUTING DISTANCE MEASURES

The distance measures are computed between the actual input speech in the sub-frame buffer and the synthesized speech segment (non-postfiltered). The procedure for computing the distance measures is exactly same as the one described in Chapter 4, but the autocorrelation lags of the input and synthesized speech segments need to be computed before calculating the distance measures.

6.5   SUMMARY

The FS-1016 CELP decoder synthesizes output speech from the transmitted parameters. The LSFs of the frame are decoded and interpolated to form sub-frame LSFs and the ACB and SCB parameters extracted from the bitstream. The extracted parameters are then used to generate a

```
% P6.3 - adaptive_postfilt.m
%    s                   - Non-postfiltered and postfiltered speech
%    alpha, beta         - Filter parameters (0.8,0.5)
%    powerin, powerout - Input and Output power estimates
%    dp1, dp2, dp3       - Filter memories
%    fci                 - LPC predictor poles
%    no                  - Predictor order
%    agcScale            - Automatic gain control scaling vector
%    TC                  - Input/Output power estimate time constant

load postfilt_dat;
% ALLOCATE LOCAL FILTER MEMORY (SPECTRAL TILT COMPENSATOR)
ast = zeros( 2, 1 );

% ESTIMATE INPUT POWER

[ newpowerin, ipZ ] = filter( [TC 0], [1 -1+TC], ( s .* s ), ipZ );
powerin = newpowerin(1);
% BANDWIDTH EXPAND PREDICTOR POLES TO FORM PERCEPTUAL POSTFILTER
pcexp1 = bwexp( beta, fci, no );
pcexp2 = bwexp( alpha, fci, no );
% APPLY POLE-ZERO POSTFILTER
[ dp1, s ] = zerofilt( pcexp1, no, dp1, s, l );
[ dp2, s ] = polefilt( pcexp2, no, dp2, s, l );

% ESTIMATE SPECTRAL TILT (1ST ORDER FIT) OF POSTFILTER
% DENOMINATOR DOMINATES (POLES)
rcexp2 = pctorc( pcexp2, no );

% ADD TILT COMPENSATION BY A SCALED ZERO (DON'T ALLOW HF ROLL-OFF)
ast(1) = 1.0;
if rcexp2(1) > 0.0
    ast(2) = -0.5 * rcexp2(1);
else
    ast(2) = 0.00;
end
[ dp3, s ] = zerofilt( ast, 1, dp3, s, l );

% ESTIMATE OUTPUT POWER
[ newpowerout, opZ ] = filter( [TC 0], [1 -1+TC], ( s .* s ), opZ )
powerout = newpowerout(1);
```

Program P6.3: Postfiltering output speech. (*Continues.*)

```
% INCORPORATE SAMPLE-BY-SAMPLE AUTOMATIC GAIN CONTROL
agcUnity = find( newpowerout <= 0 );
newpowerout( agcUnity ) = ones( length( agcUnity ), 1 );
agcScale = sqrt( newpowerin ./ newpowerout );
agcScale( agcUnity ) = ones( length( agcUnity ), 1 );
s = s .* agcScale;
```

Program P6.3: (*Continued.*) Postfiltering output speech.

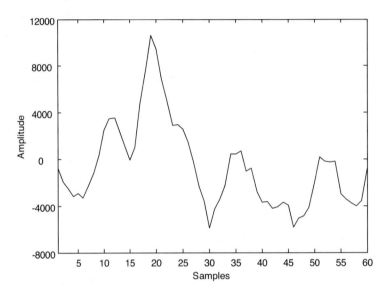

Figure 6.9: Postfiltered and scaled output speech segment.

composite excitation which is used along with the LP parameters to synthesize speech on a sub-frame basis. Since CELP uses an A-by-S procedure, the CELP encoder contains a replica of the CELP decoder except for the postfilter. The FS-1016 speech coder provides a good compromise between speech quality and bit rate, and the subjective quality of speech is maintained in the presence of moderate background noise [20]. Modern day speech coders have incorporated variable bit rate coding schemes to optimize speech quality under different channel noise conditions [5, 42]. The Selectable Mode Vocoder (SMV) classifies input speech into different categories and selects the best rate in order to ensure speech quality at a given operating mode [78].

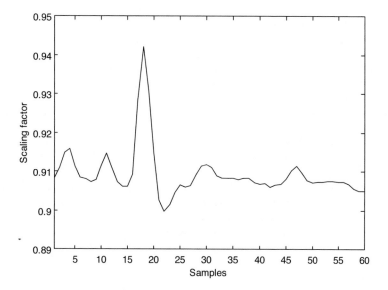

Figure 6.10: Scaling factor for postfiltered speech provided by AGC.

Bibliography

[1] M.R. Schroeder and B. Atal, "Code-Excited Linear Prediction (CELP): High Quality Speech at Very Low Bit Rates," *Proc. ICASSP-85*, p. 937, Apr. 1985. 7, 9, 63

[2] S. Singhal and B. Atal, "Improving the Performance of Multi-Pulse Coders at Low Bit Rates," *Proc. ICASSP-84*, p. 1.3.1, 1984. 7, 63

[3] B.S. Atal and J. Remde, "A New Model for LPC Excitation for Producing Natural Sounding Speech at Low Bit Rates," *Proc. IEEE ICASSP-82*, pp. 614–617, Apr. 1982. 7

[4] R. Salami *et al.*, "Design and Description of CS-ACELP: A Toll Quality 8 kb/s Speech Coder," *IEEE Trans. on Speech and Audio Proc.*, vol. 6, no. 2, pp. 116–130, Mar. 1998. DOI: 10.1109/89.661471 7, 10, 78

[5] B. Bessette *et al.*, "The Adaptive Multi-Rate Wideband Speech Codec (AMR-WB)," *IEEE Trans. on Speech and Audio Proc.*, vol. 10, no. 8, pp. 620–636, Nov. 2002. DOI: 10.1109/TSA.2002.804299 7, 11, 49, 78, 90

[6] M.R. Schroeder, B.S. Atal, and J.L. Hall, "Optimising Digital Speech Coders by Exploiting Masking Properties of the Human Ear," *J. Acoust. Soc. Am.*, vol. 66, 1979. DOI: 10.1121/1.383662 7

[7] P. Kroon, E. Deprettere, and R.J. Sluyeter, "Regular-Pulse Excitation-A Novel Approach to Effective and Efficient Multi-pulse Coding of Speech," *IEEE Trans. on Acoustics, Speech, and Signal Proc.*, vol. 34, no. 5, pp. 1054–1063, Oct. 1986. DOI: 10.1109/TASSP.1986.1164946 7

[8] I. Boyd and C. Southcott, "A Speech Codec for the Skyphone Service," *Br. Telecom Technical J.*, vol. 6(2), pp. 51–55, Apr. 1988. DOI: 10.1007/s10550-007-0070-0 7

[9] GSM 06.10, "GSM Full-Rate Transcoding," *Technical Report Version* 3.2, ETSI/GSM, Jul. 1989. 7

[10] B.S. Atal, "Predictive Coding of Speech at Low Bit Rates," *IEEE Trans. on Comm.*, vol. 30, no. 4, p. 600, Apr. 1982. 7

[11] B. Atal and M.R. Schroeder, "Stochastic Coding of Speech Signals At Very Low Bit Rates," *Proc. Int. Conf. Comm.*, pp. 1610–1613, May 1984. 7

[12] G. Davidson and A. Gersho, "Complexity Reduction Methods for Vector Excitation Coding," *Proc. IEEE ICASSP*-86, p. 3055, 1986. 9

[13] W.B. Kleijn *et al.*, "Fast Methods for the CELP Speech Coding Algorithm," *IEEE Trans. on Acoustics, Speech, and Signal Proc.*, vol. 38, no. 8, pp. 1330, Aug. 1990. DOI: 10.1109/29.57568 9

[14] I. Trancoso and B. Atal, "Efficient Search Procedures for Selecting the Optimum Innovation in Stochastic Coders," *IEEE Trans. ASSP*-38(3), p. 385, Mar. 1990. DOI: 10.1109/29.106858 9

[15] I. Gerson and M.A. Jasiuk, "Vector Sum Excited Linear Prediction (VSELP) Speech Coding at 8 kb/s," *Proc. IEEE ICASSP*-90, vol. 1, pp. 461–464, Apr. 1990. DOI: 10.1109/ICASSP.1990.115749 9, 10

[16] I. Gerson and M. Jasiuk, "Techniques for Improving the Performance of CELP-type speech coders," *Proc. IEEE ICASSP*-91, pp. 205–208, May 1991. DOI: 10.1109/49.138990 9

[17] P. Kroon and B. Atal, "Pitch Predictors with High Temporal Resolution," *Proc. IEEE ICASSP*-90, pp. 661–664, Apr. 1990. DOI: 10.1109/ICASSP.1990.115832 9, 71

[18] W.B. Kleijn, "Source-Dependent Channel Coding and its Application to CELP," *Advances in Speech Coding*, Eds. B. Atal, V. Cuperman, and A. Gersho, pp. 257–266, Kluwer Ac. Publ., 1990. 9, 12, 65

[19] A. Spanias, M. Deisher, P. Loizou and G. Lim, "Fixed-Point Implementation of the VSELP algorithm," *ASU-TRC Technical Report*, TRC-SP-ASP-9201, Jul. 1992. 9

[20] J.P. Campbell Jr., T.E. Tremain and V.C. Welch, "The Federal Standard 1016 4800 bps CELP Voice Coder," *Digital Signal Processing*, Academic Press, Vol. 1, No. 3, p. 145–155, 1991. 10, 64, 65, 66, 76, 90

[21] Federal Standard 1016, *Telecommunications: Analog to Digital Conversion of Radio Voice by 4800 bit/second Code Excited Linear Prediction (CELP)*, National Communication System - Office Technology and Standards, Feb. 1991. 10, 12

[22] TIA/EIA-PN 2398 (IS-54), "The 8 kbit/s VSELP Algorithm," 1989. 10

[23] GSM 06.60, "GSM Digital Cellular Communication Standards: Enhanced Full-Rate Transcoding," ETSI/GSM, 1996. 10

[24] GSM 06.20, "GSM Digital Cellular Communication Standards: Half Rate Speech; Half Rate Speech Transcoding," ETSI/GSM, 1996. 10

[25] ITU Draft Recommendation G.728, "Coding of Speech at 16 kbit/s using Low-Delay Code Excited Linear Prediction (LD-CELP)," 1992. 10

[26] J. Chen, R. Cox, Y. Lin, N. Jayant, and M. Melchner, "A Low-Delay CELP Coder for the CCITT 16 kb/s Speech Coding Standard," IEEE Trans. on Sel. Areas in Comm., vol. 10, no. 5, pp. 830–849, Jun. 1992. DOI: 10.1109/49.138988 10

[27] TIA/EIA/IS-96, "QCELP," Speech Service Option 3 for Wideband Spread Spectrum Digital Systems, TIA 1992. 10

[28] ITU Recommendation G.723.1, "Dual Rate Speech Coder for Multimedia Communications transmitting at 5.3 and 6.3 kb/s," Draft 1995. 10

[29] TIA/EIA/IS-641, "Cellular/PCS Radio Interface - Enhanced Full-Rate Speech Codec," TIA 1996. 10

[30] TIA/EIA/IS-127, "Enhanced Variable Rate Codec," Speech Service Option 3 for Wideband Spread Spectrum Digital Systems, TIA, 1997. 10

[31] W.B. Kleijn et al., "Generalized Analysis-by-Synthesis Coding and its Application to Pitch Prediction," Proc. IEEE ICASSP-92, vol. 1, pp. 337–340, Mar. 1992. DOI: 10.1109/ICASSP.1992.225903 10

[32] ITU Study Group 15 Draft Recommendation G.729, "Coding of Speech at 8kb/s using Conjugate-Structure Algebraic-Code-Excited Linear-Prediction (CS-ACELP)," 1995. 10

[33] J.P. Adoul et al., "Fast CELP Coding Based on Algebraic Codes," Proc. IEEE ICASSP-87, vol. 12, pp. 1957–1960, Apr. 1987. 10, 78

[34] J.I. Lee et al., "On Reducing Computational Complexity of Codebook Search in CELP Coding," IEEE Trans. on Comm., vol. 38, no. 11, pp. 1935–1937, Nov. 1990. DOI: 10.1109/26.61473 10

[35] C. Laflamme et al., "On Reducing Computational Complexity of Codebook Search in CELP Coder through the Use of Algebraic Codes," Proc. IEEE ICASSP-90, vol. 1, pp. 177–180, Apr. 1990. DOI: 10.1109/ICASSP.1990.115567 10

[36] D.N. Knisely, S. Kumar, S. Laha, and S. Navda, "Evolution of Wireless Data Services: IS-95 to CDMA2000," IEEE Comm. Mag., vol. 36, pp. 140–149, Oct. 1998. DOI: 10.1109/35.722150 11

[37] ETSI AMR Qualification Phase Documentation, 1998. 11

[38] R. Ekudden, R. Hagen, I. Johansson, and J. Svedburg, "The Adaptive Multi-Rate speech coder," Proc. IEEE Workshop on Speech Coding, pp. 117–119, Jun. 1999. 11

[39] Y. Gao et al., "The SMV Algorithm Selected by TIA and 3GPP2 for CDMA Applications," Proc. IEEE ICASSP-01, vol. 2, pp. 709–712, May 2001. DOI: 10.1109/ICASSP.2001.941013 11

[40] Y. Gao *et al.*, "EX-CELP: A Speech Coding Paradigm," *Proc. IEEE ICASSP*-01, vol. 2, pp. 689–692, May 2001. DOI: 10.1109/ICASSP.2001.941008 11

[41] TIA/EIA/IS-893, "Selectable Mode Vocoder," Service Option for Wideband Spread Spectrum Communications Systems, ver. 2.0, Dec. 2001. 11

[42] M. Jelinek and R. Salami, "Wideband Speech Coding Advances in VMR-WB Standard," *IEEE Trans. on Audio, Speech and Language Processing*, vol. 15, iss. 4, pp. 1167–1179, May 2007. DOI: 10.1109/TASL.2007.894514 11, 49, 90

[43] R. Salami, R. Lefebvre, A. Lakaniemi, K. Kontola, S. Bruhn and A. Taleb, "Extended AMR-WB for High-Quality Audio on Mobile Devices," *IEEE Transactions on Communications*, Vol. 44, No. 5, May 2006. DOI: 10.1109/MCOM.2006.1637952 12

[44] I. Varga, S. Proust and H. Taddei, "ITU-T G.729.1 Scalable Codec for New Wideband Services," *IEEE Communications Magazine*, Oct. 2009. 12

[45] D. Kemp *et al.*, "An Evaluation of 4800 bits/s Voice Coders," *Proc. ICASSP*-89, p. 200, Apr. 1989. 12

[46] D. Lin, "New Approaches to Stochastic Coding of Speech Sources at Very Low Bit Rates," *Proc. EUPISCO*-86, p. 445, 1986. 12, 65

[47] J. Chen and A. Gersho, "Real Time Vector APC Speech Coding at 4800 bps with Adaptive Postfiltering," *Proc. ICASSP*-87, pp. 2185–2188, 1987. 12, 84

[48] A. Spanias, "Speech Coding: a Tutorial Review," *Proceedings of the IEEE*, vol. 82, no. 10, pp. 1541–1582, Oct. 1994. 13

[49] A. Spanias, T. Painter and V. Atti, Audio Signal Processing and Coding, Wiley-Interscience, New Jersey, 2007. 13, 78

[50] S. Haykin, *Adaptive Filter Theory*, Prentice-Hall, New Jersey, 1996. 18, 26, 53

[51] S.L. Marple, *Digital Spectral Analysis with Applications*, Prentice Hall, New Jersey, 1987. 21

[52] T. Parsons, *Voice and Speech Processing*, McGraw-Hill, 1987. 21

[53] P. Kabal, "Ill-Conditioning and Bandwidth Expansion in Linear Prediction of Speech," *Proc. IEEE ICASSP*, vol. 1, pp. I-824-I-827, 2003. 23

[54] P.P. Vaidyanathan, *The Theory of Linear Prediction*, Morgan & Claypool, 2008. DOI: 10.2200/S00086ED1V01Y200712SPR003 27

[55] J. Makhoul, "Linear Prediction: A Tutorial Review," *Proc. IEEE*, Vol. 63, No. 4, pp. 561–580, Apr. 1975. DOI: 10.1109/PROC.1975.9792 28

[56] J. Markel and A. Gray, Jr., *Linear Prediction of Speech*, Springer-Verlag, New York, 1976. 28

[57] T.E. Tremain, "The Government Standard Linear Predictive Coding Algorithm: LPC-10," *Speech Technology*, pp. 40–49, Apr. 1982. 28

[58] G. Box and G. Jenkins, *Time Series Analysis Forecasting and Control*, HoldenDay, Inc., San Francisco, 1970. 28

[59] A. Gray and D. Wong, "The Burg Algorithm for LPC Speech Analysis/Synthesis," *IEEE Trans. ASSP*-28, No. 6, pp. 609–615, Dec. 1980. 28

[60] V. Viswanathan and J. Makhoul, "Quantization Properties of Transmission Parameters in Linear Predictive Systems," *IEEE Trans. ASSP*-23, pp. 309–321, Jun. 1975. 28

[61] N. Sugamura and F. Itakura, "Speech Data Compression by LSP Analysis/Synthesis Technique," *Trans. IEICE*, Vol. J64, pp. 599–606, 1981. 29

[62] F.K. Soong and B-H Juang, "Line Spectrum Pair (LSP) and Speech Data Compression," *Proc. IEEE ICASSP*, pp. 1.10.1–1.10.4, Mar. 1984. 29

[63] P. Kabal and R. Ramachandran, "The Computation of Line Spectral Frequencies using Chebyshev Polynomials," *IEEE Trans. ASSP*-34, No. 6, pp. 1419–1426, Dec. 1986. 49, 52

[64] G. Kang and L. Fransen, "Application of Line-Spectrum Pairs to Low-Bit-Rate Speech Encoders," *Proc. IEEE ICASSP*, vol. 10, pp. 244–247, 1985. 49

[65] J.S. Collura and T.E. Tremain, "Vector Quantizer Design for the Coding of LSF Parameters," *Proc. IEEE ICASSP*, vol. 2, pp. 29–32, Apr. 1993. 49

[66] N. Farvardin and R. Laroia, "Efficient Encoding of Speech LSP Parameters using the Discrete Cosine Transformation," *Proc. IEEE ICASSP*, vol. 1, pp. 168–171, May 1989. DOI: 10.1109/ICASSP.1989.266390 49

[67] K.K. Paliwal and B.S. Atal, "Efficient Vector Quantization of LPC Parameters at 24 bits/frame," *IEEE Trans. on Speech and Audio Processing*, vol. 1, iss. 1, pp. 3–14, Jan. 1993. DOI: 10.1109/89.221363 49

[68] J. Pan and T.R. Fischer, "Vector Quantization of Speech Line Spectrum Pair Parameters and Reflection Coefficients," *IEEE Trans. on Speech and Audio Processing*, vol. 6, iss. 2, Mar. 1998. DOI: 10.1109/89.661470 49

[69] Y. Bistritz and S. Peller, "Immittance Spectral Pairs (ISP) for Speech Encoding," *Proc. IEEE ICASSP*, vol. 2, pp. 9–12, Apr. 1993. DOI: 10.1109/ICASSP.1993.319215 49

[70] A. Gray Jr. and J. Markel, "Distance Measures for Speech Processing," *IEEE Trans. ASSP*, vol. 24, pp. 380–391, Oct. 1976. 57, 60

[71] F. Itakura, "Minimum Prediction Residual Principle Applied to Speech Recognition," *IEEE Trans. ASSP*, vol. 23, pp. 67–12, Feb. 1975. DOI: 10.1109/TASSP.1975.1162641 60

[72] A.V. Oppenheim, R.W. Schafer and T.C. Stockham, "Non-linear Filtering of Multiplied and Convolved Signals," *Proc. IEEE*, vol. 56, pp. 1264–1291, Aug. 1968. DOI: 10.1109/PROC.1968.6570 60

[73] A. Gersho and V. Cuperman, "Vector Quantization: A Pattern Matching Technique for Speech Coding," *IEEE Com. Mag.*, Vol. 21, p. 15, Dec. 1983. 63

[74] R. Gray, "Vector Quantization," *ASSP Mag.*, Vol. 1, p. 4, Apr. 1984. 63

[75] M. Sabin and R. Gray, "Product Code Vector Quantizers for Speech Waveform Coding," *Proc. Globecom*-82, p. E6.5.1, 1982. 64

[76] Website for the book. Available online at: http://www.morganclaypool.com/page/fs1016 60, 68, 70, 72, 75

[77] S. Lin and D.J. Costello, Jr., *Error Control Coding*, Prentice Hall, New Jersey, 2004. 77, 80

[78] S.C. Greer and A. DeJaco, "Standardization of the Selectable Mode Vocoder," *Proc. IEEE ICASSP*, vol. 2, pp. 953–956, May 2001. DOI: 10.1109/ICASSP.2001.941074 90

Authors' Biographies

KARTHIKEYAN N. RAMAMURTHY

Karthikeyan N. Ramamurthy completed his MS in Electrical Engineering and is currently a PhD student in the School of Electrical, Computer, and Energy Engineering at Arizona State University. His research interests include DSP, speech processing, sparse representations and compressive sensing. He has worked extensively with the CELP algorithm and has created MATLAB modules for the various functions in the algorithm. He also works in signal processing and spectral analysis for Earth and geology systems and has created several functions in Java-DSP software for the same.

ANDREAS SPANIAS

Andreas Spanias is a Professor in the School of Electrical, Computer, and Energy Engineering at Arizona State University (ASU). He is also the founder and director of the SenSIP industry consortium. His research interests are in the areas of adaptive signal processing, speech processing, and audio sensing. He and his student team developed the computer simulation software Java-DSP (J-DSP – ISBN 0-9724984-0-0). He is author of two text books: *Audio Processing and Coding* by Wiley and *DSP: An Interactive Approach*. He served as Associate Editor of the IEEE Transactions on Signal Processing and as General Co-chair of IEEE ICASSP-99. He also served as the IEEE Signal Processing Vice-President for Conferences. Andreas Spanias is co-recipient of the 2002 IEEE Donald G. Fink paper prize award and was elected Fellow of the IEEE in 2003. He served as Distinguished lecturer for the IEEE Signal processing society in 2004.

Printed in the United States
by Baker & Taylor Publisher Services